T0271006

Cloud-based Multi-Modal Information Analytics

Cloud-based Multi-Modal Information Analytics: A Hands-on Approach discusses the various modalities of data and provides an aggregated solution using cloud. It includes the fundamentals of neural networks, different types and how they can be used for multi-modal information analytics. The various application areas that are image-centric and videos are also presented with deployment solutions in the cloud.

Features

- Lifecycle of multi-modal data analytics is discussed with applications of modalities of text, image and video
- Deep learning fundamentals and architectures covering convolutional neural networks, recurrent neural networks and types of learning for different multi-modal networks
- Applications of multi-modal analytics covering text, speech and image

This book is aimed at researchers in multi-modal analytics and related areas.

Chapman & Hall/CRC Cloud Computing for Society 5.0

Series Editor: Vishal Bhatnagar and Vikram Bali

Digitalization of Higher Education using Cloud Computing
Edited by: S.L. Gupta, Nawal Kishor, Niraj Mishra, Sonali Mathur, and Utkarsh Gupta

Cloud Computing Technologies for Smart Agriculture and Healthcare
Edited by: Urmila Shrawankar, Latesh Malik, and Sandhya Arora

Cloud and Fog Computing Platforms for Internet of Things
Edited by: Pankaj Bhambri, Sita Rani, Gaurav Gupta, and Alex Khang

Cloud based Intelligent Informative Engineering for Society 5.0
Edited by: Kaushal Kishor, Neetesh Saxena, and Dilkeshwar Pandey

Cloud-based Multi-Modal Information Analytics: A Hands-on Approach
Authors: Srinidhi Hiriyannaiah, Siddesh G M, and Srinivasa K G

For more information about this series please visit: https://www.routledge.com/ Chapman--HallCRC-Cloud-Computing-for-Society-50/book-series/CRCCCS

Cloud-based Multi-Modal Information Analytics

A Hands-on Approach

Srinidhi Hiriyannaiah
Siddesh G M
Srinivasa K G

CRC Press
Taylor & Francis Group
Boca Raton London New York

CRC Press is an imprint of the
Taylor & Francis Group, an **informa** business

A CHAPMAN & HALL BOOK

First edition published 2023
by CRC Press
6000 Broken Sound Parkway NW, Suite 300, Boca Raton, FL 33487-2742

and by CRC Press
4 Park Square, Milton Park, Abingdon, Oxon, OX14 4RN

CRC Press is an imprint of Taylor & Francis Group, LLC

Library of Congress Cataloging-in-Publication Data
Names: Hiriyannaiah, Srinidhi, author. | Siddesh, G. M., 1981- author. |
Srinivasa, K. G., author.
Title: Cloud based multi-modal information analytics : a hands-on approach
/ Srinidhi Hiriyannaiah, Siddesh G M, Srinivasa K G.
Description: First edition. | Boca Raton : Chapman & Hall/CRC Press, 2023.
| Series: Chapman & Hall/CRC cloud computing for society 5.0 | Includes
bibliographical references. | Summary: "Cloud based Multi-Modal
Information Analytics: A Hands-on Approach discusses the various
modalities of data and provide an aggregated solutions using cloud. It
includes the fundamentals of neural networks, different types and how it
can be used for the multi-modal information analytics. The various
application areas that are image-centric and video are also presented
with deployment solutions in the cloud"-- Provided by publisher.
Identifiers: LCCN 2022055693 (print) | LCCN 2022055694 (ebook) | ISBN
9781032105673 (hardback) | ISBN 9781032493138 (paperback) | ISBN
9781003215974 (ebook)
Subjects: LCSH: Quantitative research--Data processing. | Cloud computing.
Classification: LCC QA76.9.Q36 H57 2023 (print) | LCC QA76.9.Q36 (ebook)
| DDC 001.4/2--dc23/eng/20230123
LC record available at https://lccn.loc.gov/2022055693
LC ebook record available at https://lccn.loc.gov/2022055694

ISBN: 978-1-032-10567-3 (hbk)
ISBN: 978-1-032-49313-8 (pbk)
ISBN: 978-1-003-21597-4 (ebk)

DOI: 10.1201/9781003215974

Typeset in Palatino
by SPi Technologies India Pvt Ltd (Straive)

Contents

Preface

With the development of web technology, multi-modal or multi-view data has surged as a major stream for big data, where each modal/view encodes individual property of data objects. Often, different modalities are complementary to each other. Such a fact motivated a lot of research attention to fusing the multi-modal feature spaces to comprehensively characterize the data objects. Most of the existing state-of-the-art focuses on how to fuse the energy or information from multi-modal spaces to deliver superior performance over their counterparts with a single modal. Recently, deep neural networks have been exhibited as a powerful architecture to well-capture the nonlinear distribution of high-dimensional multimedia data, so naturally does for multi-modal data In multimedia systems, it is important how to integrate different modalities from the data in order to achieve maximum performance and harvest relevant information. Traditional data fusion techniques usually include early fusion, late fusion and middle fusion. It is widely known that deep learning has become a powerful toolkit in the field of pattern recognition that largely improved the performance of the traditional multi-modal models, yielding the so-called deep multi-view/modal models, which can effectively extract the high-level nonlinear feature representation from the multi-modal data to benefit the tasks.

The aim of this book is to present different dimensions of multi-modal and deep learning methods using three different modalities, including video, audio and image information. The solutions presented leverage both spatial and temporal information from multi-modal data and effectively integrate them for interpretation and analysis. The book is divided into three parts and ten chapters. Part I discusses the introduction to multi-modal data and analytics that describes various modalities of data. Subsequently, Part II highlights the different architectures used in analytics of multi-modal data. After that, Part III provides various application-centric examples of different modalities, including video, audio and image. This book also provides a platform for most recent research on using deep learning-based solutions for multi-modal data analytics. The book is logically divided into three parts. The first part deals with the gentle introduction to cloud-based multi-model data analytics, the second part provides an architecture and suitable examples for multi-modal data and analytics using cloud, and, finally, the third part explores different cloud-based applications that require multi-modal analytics.

Organization of the Book

Chapter 1 provides an overview of multi-modal data analytics and life-cycle of development of an application using cloud-based utilities. It introduces the various types of multi-modal data and their applications and challenges of multi-modal data analytics.

Chapter 2 explores the different Google Cloud services, storage and computer engine. It also briefs how to work with Google Colaboratory.

Chapter 3 provides an overview of deep learning.

Chapter 4 focuses on deep learning platforms like OpenCV, PyTorch, TensorFlow and Keras.

Chapter 5 discusses the use of neural network models like CNN, RNN, LSTM and GRU for multi-modal data analytics.

Chapter 6 provides illustrative examples of neural networks multi-modal architectures like AlexNet, VGG-16 and YoloV3.

Chapter 7 presents a step-by-step procedure to be adopted for training neural networks on cloud, including use of distributed training, setting up of hyperparameters and optimization.

Chapter 8 provides a classical example of image analytics using Google Cloud.

Chapter 9 explores yet another classical example of text analytics via Google Cloud.

Chapter 10 concludes the book by exploring the deployment of speech analytics on Google Cloud.

The book is well-researched and written comprehensively and compellingly, making this book a must-read for all students, professionals and researchers in the field of data science. It serves as a practical guide to multi-modal information analytics through Google Cloud based on technological trends and real-world applications.

Key Features of the Book

- Clearly provides a basic understanding of the evolving field of multi-modal data analytics
- Provides excellent content on deployment of information analytics on Google Cloud with various real-life applications and use cases
- Gives practical hands-on approach with concepts and applications using Google Cloud
- Analyzes various data and data usage formats to ensure that you use the data you gather to improve analytics outcome
- Provides content that applies not only to readers but also to trainers and practitioners who want to build analytics capability beyond traditional data analytics, too

This book is really intended for readers who have no prior knowledge in data analytics. The book functions as an introductory text to multi-modal information analytics for those who want to do something beyond traditional data analytics.

Srinidhi Hiriyannaiah
Siddesh G M
Srinivasa K G

Acknowledgment

We attribute our efforts in completing this book to all the people who have inspired us and shaped our careers. We thank our college management, colleagues and students who encouraged us to work on this book.

The authors would like to thank Dr. NVR Naidu, Principal, Ramaiah Institute of Technology, for his valuable guidance and support. The authors would like to acknowledge valuable inputs and suggestions by colleagues of the departments of CSE and ISE, Ramaiah Institute of Technology.

We would like to thank Bhawansh Narain Saxena, Sanjay Raghavendra, Aman Malali and Rushali Mohbe for their contribution and support toward code implementation as a part of this book.

Srinivasa would like to thank Dr. Pradeep Kumar Sinha, Vice Chancellor and Director, IIIT Naya Raipur, for his valuable suggestions and continuous encouragement.

We are extremely grateful to our families, who graciously accepted our inability to attend to family chores during the course of writing this book, and especially for their extended warmth and encouragement. Without their support, we would not have been able to venture into writing this book.

We acknowledge the Google Cloud Platform that can be used for educational purposes that are adopted in the book for discussing the different concepts of data analytics.

Last, but not the least, we express our heartfelt thanks to the editorial team at the CRC Press, who guided us through this project.

Authors

Srinidhi Hiriyannaiah works as a Senior Software Engineer in GE Healthcare. He received his Ph.D. degree from VTU in 2020 and did his Master of Technology in Software Engineering from M.S. Ramaiah Institute of Technology, Bengaluru (VTU). He worked as an Assistant Professor in the Department of Computer Science and Engineering at M.S. Ramaiah Institute of Technology, Bengaluru, from 2016 to 2022. He previously worked at IBM India Software Labs, Bengaluru. His main area of interest includes studies related to parallel computing, big data and its applications, information management and software engineering for education.

Siddesh G M currently works as a Professor in the Department of Computer Science and Engineering (Cyber Security), M.S. Ramaiah Institute of Technology, Bangalore. He has published a good number of research papers in reputed international conferences and journals. He has authored books on Network Data Analytics, Statistical Programming in R and Internet of Things with Springer, Oxford University Press and Cengage publishers, respectively. He has edited research monographs in the area of Cyber-Physical Systems, Fog Computing and Energy Aware Computing, and Bioinformatics with CRC Press, IGI Global and Springer publishers, respectively. His research interests include Data Science, Cloud Computing, Internet of Things etc.

Srinivasa K G is a Professor of Data Science and Artificial Intelligence Programme at DSPM IIIT Naya Raipur, C. G. India. Earlier he worked as a Professor at Information Management and Emerging Engineering Department of the National Institute of Technical Teachers Training and Research, Chandigarh, an autonomous Institute under the Ministry of Education, Government of India. He also worked as an Associate Professor at CBP Government Engineering College, New Delhi (through UPSC) between 2016 and 2019. He also served as a Professor in the Department of Computer Science and Engineering at M.S. Ramaiah Institute of Technology, Bangalore, between 2003 and 2016. He received his Ph.D. in Computer Science and Engineering from Bangalore University in 2007. He is the recipient of the All India Council for Technical Education – Career Award for Young Teachers, Indian Society of Technical Education – ISGITS National Award for Best Research Work Done by Young Teachers, Institution of Engineers (India) – IEI Young Engineer Award in Computer Engineering, Rajarambapu Patil National Award for Promising Engineering Teacher Award from ISTE – 2012, and IMS Singapore – Visiting Scientist Fellowship Award. He has published

more than 150 research papers in international conferences and jour-
nals. He has visited many universities abroad as a visiting researcher: the
University of Oklahoma, USA; Iowa State University, USA; Hong Kong
University; Korean University; National University of Singapore; and the
University of British Columbia, Canada. He has authored many books in
the area of Learning Analytics, Network Data Analytics, Soft Computing,
Social Network Analysis, High-Performance Computing, R Programming,
etc., with prestigious international publishers like Springer, TMH, Oxford,
Cengage and IGI Global. He has edited research monographs in the area of
Cyber-Physical Systems, and Fog Computing and Energy Aware Computing
with CRC Press and IGI Global. He has been awarded BOYSCAST Fellowship
by DST, Government of India, for conducting post-doctoral research work at
the University of Melbourne, Australia. He is the principal Investigator for
many funded projects from AICTE, UGC, DRDO and DST. He has under-
taken consultancy projects worth 60 lakhs toward conducting Professional
Development Programmes under World Bank Project. He is a senior mem-
ber of IEEE and ACM. His recent research areas include Innovative Teaching
Practices in Engineering Education, pedagogy, outcome-based education
and teaching philosophy.

Part I

Introduction to Cloud-based Multi-Modal Data and Analytics

1

Multi-Modal Data Analytics and Lifecycle using Cloud

1.1 Introduction

There are various modalities with valuable information in real-world applications. It is essential to analyze the information in such modalities using learning algorithms and enhance the applications. The majority of the applications focus on single modality rather than the other modalities that exist in the applications. For example, emotion recognition applications most often use the face (images) as the only modality but not the other modality of speech. The consideration of different modalities will enhance the applications.

The advancements in the web and the internet have paved the way for big data in many areas of applications ranging from retail to cloud services. Big data involves different modalities such as image, video, audio and text as the major forms of data analytics. Most often big data uses the text as the major single modality for analysis. The image and video modality is taken as another dimension with a separate entity for data analytics. However, the modalities of data can complement each other in some data analytical applications [1]. For example, in emotion recognition, image and video complement each other for analytics. This has motivated the use of fusion of multiple modalities for data analytical applications. The existing state-of-the-art works exhibit the fusion of different modalities for the development of superior performance-based applications. Most of the works use deep learning models as the base work for analytics. Deep learning models have powerful architectures that capture high-dimensional data with multi-modalities in a non-linear way. There are substantial empirical evidences and applications where the different multi-modal methods have benefited.

The exponential advances in the web have doubled the amount of data through social media, blogs, communities and other sharing information sites. Most of this data falls into three major modalities: image, text and video. The huge data of the web has prompted insights into user data for recommendations, products, services and any other remarks [2, 3]. The information and the services provided are in the applications of health, education, tourism, e-commerce and others. The physiological transfer of data also exists in such modalities as healthcare to insurers and social media to e-commerce. The characteristics of the multi-modalities are volume, variety and veracity, which is often called big data. The applications of big data in business and industry expand a wider horizon in the fields of customer analytics, healthcare systems, online shopping and others. The advancements in the computing fields of deep learning, machine learning, computer vision and cloud services have paved the way to gather such data and insights into the user data.

The focus of this chapter is to introduce the different modalities and life-cycle of the applications dealing with it. It also introduces the different levels of fusion that can be achieved with multi-modalities. The chapter concludes with the different cloud applications of multi-modal analytics.

1.2 Multi-Modal Levels

The multi-modalities of data and its level of fusion are categorized into two major levels, namely early fusion/feature and late fusion/decision. In this section, we describe these levels of fusion, structure and related examples [4].

1.2.1 Feature-Level Fusion

The multi-modality-based applications act upon the features present in the data sources. The feature-level fusion approach extracts various features from the input data, which are then merged in an analysis unit (AU) for analysis. The AU performs the main analysis part of extraction and merging of various features as shown in Figure 1.1. F_1, F_2 ... F_n represent the different features in the multi-modal streams. The AU outputs the final feature as F_f combining all the features. For example, in emotion recognition applications, the various features of the face, lips, eyes and chin are taken together for the final emotion. A number of features that need attention for the feature-level fusion are discussed as follows [5].

- **Visual features**: The visual features are based on texture, color, shape, angle, etc. The visual features are extracted often from image or video. The segments of features are extracted from the image in the form of blocks or fixed-sized patches.

- **Text features**: These are based on caption, transcript, speech, etc. The textual features can be extracted from the speech using an audio library. The OCR features can be extracted within a character.

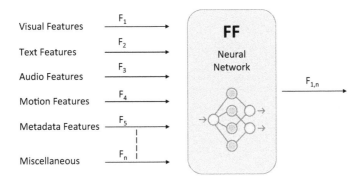

FIGURE 1.1
Feature-level fusion.

- **Audio features**: These features are based on fast Fourier transform (FFT), mel-frequency cepstral coefficient (MFCC), linear predictive coding (LPC) and others. The features of audio are often from the speech data with different types of frequencies.
- **Motion features**: These features are based on the variation of pixel measures within a frame image, histogram, motion and other patterns. These features are captured more often in the images/video.
- **Metadata features**: The supplementary features of the different modalities such as timestamp, source location and duration need to be also taken into account for the analysis. These features are extracted based on the availability and the source of the applications.

1.2.2 Decision-Level Multi-Modal Fusion

The analysis of the features is based on the decisions D_1, D_2, D_3,..., D_n based on the individual features F_1, F_2, F_3,..., F_n. The decisions are then combined using a Decision Unit (DF unit) to provide the final decision based on a hypothesis as shown in Figure 1.2. The DF combines the decisions based on conditions/objectives of the application. For example, in an emotion recognition application, D_1 can be based on the face data and D_2 can be based on the audio. The final decision of the emotion is based on the combined decisions D_1 and D_2.

The decision-level fusion provides an advantage over feature-level fusion as the modalities of data have different representations. The decision unit provides the same representation, which is the combination of all the decisions. It also offers better scalability and flexible fusion as the local decisions are involved in the decision-fusion process. However, the correlation of the

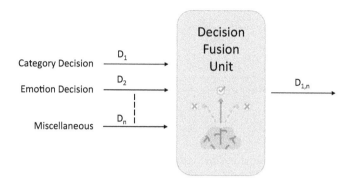

FIGURE 1.2
Decision-level fusion.

features will be missing in the decision-level fusion approaches as the decision unit solely considers only the decisions and not the correlated features.

In this section, we discussed the major approaches for multi-modal fusion, namely decision fusion and feature fusion. It discussed how the different features and decisions can be combined in a fusion manner depending on the applications. However, there are certain methods/techniques to achieve the fusion. The different methods of multi-modal fusion are discussed in the next section.

1.2.2.1 Methods for Multi-Modal Fusion

The different methods of multi-modal fusion are based on machine learning and other classification methods. In this section, we discuss various multi-modal fusion methods for the analysis tasks. The multi-modal fusion methods are classified into three major categories, namely rule-based, classification-based and estimation-based methods, as shown in Figure 1.3 [6]. In this section, we describe each method with an example of its application.

1.2.3 Rule-Based Fusion Methods

The fusion methods are based on the rules for combining various multi-modal information. The basic statistics of the fusion include sum, maximum, minimum, or, and methods. The combination of these rules is mapped based on the applications and customization rules. Examples of rule-based fusion methods include speech recognition, video retrieval, image processing, face recognition and monologue detection. The rule-based methods depend on the domain so that necessary operations can be used in an application. We describe some of the rule-based fusion methods as follows.

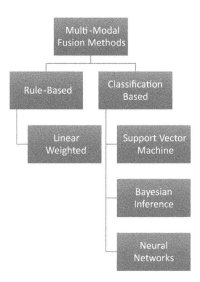

FIGURE 1.3
Multi-modal fusion types.

1.2.4 Linear Weighted Fusion

It is one of the simple methods of fusion where the information of different modalities is combined in a linear fashion. The information can be based on audio features, visual features, speech features or other semantic-level decisions [7]. The combination of the information needs to start with the normalized weights in different modalities. There are various methods for the normalization of weights, namely min-max, z-score, tanh and sigmoid estimators. Most often, methods used for the normalization are min-max and z-score normalization.

The methodology for the linear weighted fusion is described using Equations 1.1 and 1.2. Initially, each input feature I_i is obtained from the ith modality source. A normalized weight w_i is assigned to each input feature I_i to obtain the normalized input feature I_N as shown in Equation 1.1. The linear fusion technique of sum/max/min/product is then applied to I_N to obtain the fusion modality. This technique of linear fusion is applied in applications like video surveillance and traffic monitoring, detecting and tracking people and other multimedia analysis tasks.

$$I_N = \sum_{i=1}^{n} w_i * I_i \qquad (1.1)$$

$$I_{f=} \prod_{i=1}^{n} I_n \qquad (1.2)$$

1.2.5 Classification-Based Fusion Methods

There are numerous methods used for multi-modal decision fusion into pre-defined classes. These methods primarily focus on the aspects of decision fusion into a pre-defined class based on the local decisions and the features. Some of the methods used often are support vector machine (SVM), Bayesian inference, neural networks, cross-entropy model and others. The different methods used for classification are described as follows.

1.2.5.1 Support Vector Machine

It is one of the popular methods used for classification and other related tasks. For multi-modal data and classification, SVM is used for text categorization, face recognition, concept classification, etc. It is based on the supervised learning approach, wherein the input data is partitioned based on the line of one or two learned classes. The multi-modal fusion with SVM uses the individual feature/decision scores for the final classification. The SVM involving text, video and audio with individual scores was explored for multi-modal classification [8]. The individual scores from each modality are used to create the final high-dimensional vector for the classification. In this way, SVM can be used for multi-modal classification.

1.2.5.2 Bayesian Inference

It is an often-used method for multi-modal fusion, which is based on the rules of probability. The inference decision can be made either with the feature or decision [9]. A joint probability is derived based on the features and decisions for each modality and its score as shown in the equation. Here for a hypothesis 'H' the posterior probabilities are estimated for all the 'n' classes using a normalized weight 'w'. Then the decision of the fusion is established using the maximum probability rule as shown in the equation. In this way, the decision fusion is established for the given modalities of the data.

$$p(H|I_1, I_2, I_3, \ldots, I_n) = \frac{1}{N} \prod_{j=1}^{n} p(I_j|H)^{w^j} \qquad (1.3)$$

$$\hat{H} = \max\left(p\left(H|I_1, I_2, I_3, \ldots, I_n\right)\right) \qquad (1.4)$$

1.2.5.3 Neural Networks

Neural networks are considered one of the black-box methods to solve expensive problems. It mainly consists of three elements: input, hidden and output nodes. The network architecture using these three elements forms the basis of the decision of the fusion [10]. The interconnections between the nodes and the weights assigned to them also influence the inferences made over for the fusion. Neural networks can be applied either to the feature or decision level depending on the applications as shown in Figure 1.4. The different types of modalities are analyzed separately using different neural networks. It is then combined to form a high-dimensional vector for the final classification using a fusion algorithm.

The Bayesian inference and the SVM methods are often used for multi-modal decision fusion inference. However, both of these methods rely on the labeling of the training data for the test data classification as it is supervised learning. Neural networks on the other side support the non-linear way classification. The focus of this book is on neural networks and their variants for multi-modal fusion. In the upcoming sections of the book, we first discuss the types of multi-modal data and the related cloud applications.

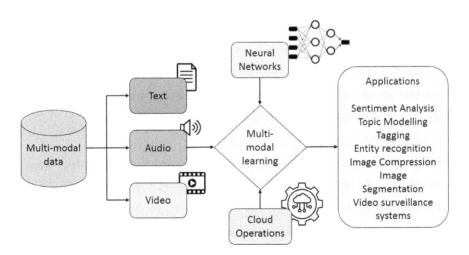

FIGURE 1.4
Multi-modal data analytics.

1.3 Types of Multi-Modal Data

Multi-modal data include image, audio, video and text. The fast and wide-spread of the internet and cloud have resulted in various big data systems. Multimedia data-sharing websites like YouTube, Instagram, Facebook and other social networks have been considered the major sources of multi-modal data. Multi-modal data analytics consists of managing, manipulating and visualizing the data effectively. It includes the applications of computer vision, database management, speech processing and recognition, video processing in various domains like education, healthcare and entertainment [11]. In this section, we describe the various types of multi-modal data and their applications as shown in Figure 1.4.

1.3.1 Text

The basic form of the data in various analytical applications is in the form of text. It may be present in the form of web pages, social network feeds, surveys and metadata. It may be also embedded in structured or unstructured formats. Extract-transform-load (ETL) tools can be used for processing of textual information in databases and other systems [12]. However, some applications have text data in an unstructured format like tweets, blogs, etc. These require attention of developing modules that are needed for extraction and analysis in the form of text analytics. There are various text analytics and methods for extraction of text and conversion into structured format. The use of machine learning and deep learning methods has been very helpful in various applications. The rule-based decision-making process is being evolved into an intelligent action-based agent. The different types of applications based on text analytics are discussed as follows.

- **Sentiment analysis:** It identifies the different types of emotions present in the text. It helps in determining the polarity of a product or an issue. It may be used for social media analysis, surveys and other product issues.
- **Topic modeling:** It identifies and groups various topics of text into various categories. The text analytics of different types of categories helps in identification of domains.
- **Tagging:** Text analytics and classification are also used for tagging applications. The tagging of online articles into categories of healthcare, technology, engineering and others helps students, professors and scientists to gather knowledge based on a particular topic.
- **Entity recognition and extraction:** Entities refer to the places, persons and locations that are present in the text form. Text analytics that involves the extraction of specific parts of the text and identifies the entities help in various applications.

Text analytics, combined with applications, also faces various challenges. It includes complexity, conceptual and contextual struggles. The complexity of transformation of the text into a specific format is challenging and needs infrastructure. It includes collection of data, keyword detection, class definition and grouping. The contextual understanding of text is of utmost importance for text analytics. The programmers have to find an effective way for organizing and construct meaningful sentences for analytics. Text data is often used in various industrial applications in robotics. However, we most often find images and videos as the data shared in most social networks. Hence, there is a need for better understanding of these multi-modal data as well. The image and video data are discussed in the next sections.

1.3.2 Image

Images are an important form of data that is used for different applications. The advances in computing have enabled the storage and analysis of images in an efficient way. Computer vision with images has paved the way for different types of applications in different domains. Image processing is mainly used in medical and healthcare for the diagnosis and classification of diseases. The use of deep learning models and competitive challenges has paved the way for numerous architectures for image classification. For example, ImageNet challenge [13] enabled the solutions of AlexNet, GoogleNet, VGG, ResNet, etc., which are used as the baselines for image classification in different applications. These architectures promised the classification of images using CNNs with a varied number of layers. The different types of applications that are based on image analytics are listed as follows.

- **Image compression and decompression:** The size of the images and their variants play a role in image analytics. The compression and decompression of the images help in the storage and analysis for the applications in healthcare. The compression can be lossless or lossy. Lossless compression is preferred for medical applications, clip arts or technical drawings. Lossy methods are used for the photographs with substantial decrease in bit rate.

- **Image noising and denoising:** These applications are used for the images for variation in color or brightness. Image noise refers to the external signals present in the image that results in the degradation of the images. Denoising involves the reconstruction of the image by removing the noises in the images.

- **Image segmentation:** It involves the mechanism of partitioning of the image into multiple segments for transformation and representation change. It is mainly used for locating objects, boundaries and curves in images. It involves the process of assigning the label and sharing the characteristics.

- **Image synthesis and classification:** Image synthesis involves the generation of images in large quantities with wide variety and distribution. The synthetic data generation and analysis are required for image applications where the data is unknown. For example, image synthesis and generation are used for the text-to-image data.

1.3.3 Visual and Audio

Visual data: Visual data, including images and videos (image sequences), are the most common and challenging multimedia data due to their rich information and semantic contents that form almost 80% of all unstructured big data (Venter and Stein 2012). Video and image analytics is the process of extracting meaningful concepts and information from unstructured visual data. The main challenge of visual data is their huge size compared to structured or textual data, which is why big data solutions are being used. In recent years, visual data have been generated exponentially due to mobile technologies, high-performance cloud computing, and low-cost storage and sharing websites. Video surveillance systems, autonomous vehicles, video and image retrieval systems, and healthcare are a few applications of visual data analytics. In particular, although the human brain is able to efficiently analyze millions of visual data extracted from various sources in parallel, advanced technology is beating human brain performance specifically in visual data. Microsoft, for instance, beat humans at the ImageNet Large Scale Visual Recognition Competition in 2015 [14]. This is the first time a machine outperforms humans in image classification by mimicking brain functions.

    Audio data: Another data type widely seen in multimedia applications is audio or speech data. Social media, industrial machines and medical devices are a few examples of big data sources that need real-time audio analytics. With the rapid growth of mobile technologies and applications, as well as producing longer battery life and faster processors, we are now capable of analyzing thousands of different acoustic data, including music, speech and bodily sounds in an efficient manner. Audio analytics refers to the procedure of retrieving meaningful information from unstructured aural data. For example, wearable technology is a powerful technique to capture and analyze various sounds of human body from the heart rate to the breathing sound and the digestive system. Call centers are another example where audio and speech analytics are used to recognize spoken words from thousands of calls and to improve the interaction with customers. In all the examples, big data techniques such as Hadoop and Spark have been utilized to efficiently mine unstructured data and get a better understanding of users' needs. The different types of applications that are based on the video and audio data are listed as follows.

Video surveillance systems: In video surveillance systems, the intent of the observation for the improper behavior based on the scenes is captured. These systems are used the common public places for safety and security [15]. It includes the identification of the locations and camera-specific areas for monitoring. Improper activity can be monitored based on the specific schedules and images that are appropriate for the analysis. The various applications that are used as a part of video surveillance systems are remote video monitoring, facility operations, facial recognition, traffic monitoring and public safety.

Autonomous vehicles: The automotive industry is evolving with computer vision and its applied areas toward autonomous vehicles [16]. An autonomous vehicle does not require human intervention for the operation. The operation is based on the various sensors around the vehicle that capture the video and guide the vehicle appropriately. The multi-modality of both image and video are combined for the operations of autonomous vehicles.

Emotion recognition: One of the advanced applications based on video data is emotion recognition [17]. For example, the video data of a class environment can be captured to see the different emotions of delivery like satisfaction or non-satisfaction. It can also be extended to speech-based applications for the classification of different types of emotions like happiness, anger and sadness. The speech emotion classification deals with the audio data, and thus multi-modal analytics will play a major role further.

1.4 Cloud Applications of Multi-Modal Analytics

Recent advances in artificial intelligence and machine learning have led to the development of various applications in the areas of healthcare, speech recognition, image captioning, gesture recognition and others. There are various intelligent algorithms behind these applications for pattern recognition and mining. Most often the algorithms used are clustering, Bayesian, SVM and others. However, these methods are restricted by the usage of the hardware resources on the scalability of the data. Deep learning methods have paved the way for advanced data applications through the usage of GPU. The extraction and synthesis of information in the multi-modal fusion approach require the use of GPU and other hardware resources. Deep learning has been used in several applications related to computer vision and other disciplines. It has demonstrated potential in wide areas like NLP, speech recognition, machine translation and other fields. It provides the ability to deal with the multi-modalities of information from different sources. In this section, we provide an overview of different multi-modal applications in different areas.

Figure 1.5. gives the classification of different multi-modal applications and challenges.

Healthcare

The rapidly growing population and aging population have posed various challenges in healthcare with diagnosis and data maintenance [18]. Healthcare applications have to deal with different sources of data and cannot rely on a single source. Most often rule-based reasoning systems are used for providing various decisions in the healthcare systems. However, rule-based reasoning affects the system performance and becomes unsatisfactory with some loss of data. Multi-modal systems play a key role in healthcare as the different sources of data can be combined to provide a fusion decision. In this regard, some attempts have been made in healthcare for multi-modal analytics.

Human recognition

Human recognition applications are increasing day by day due to the advances in surveillance and monitoring systems. It seeks to identify the

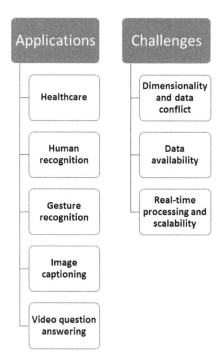

FIGURE 1.5
Multi-modal applications and challenges.

person with some labeling mechanisms using videos as the source [19]. However, other modalities like audio can also be taken into account for human recognition applications. There are recent works that focus on human recognition applications.

Gesture recognition

Gesture recognition applications are applied in the human–computer interaction research area to detect motion and actions in videos [20]. It facilitates the detection of various gestures like walking, running and sleeping based on the images/video. These applications deal with multiple sources of data like video/images. Recent works on gesture recognition include driver assistance, smart surveillance, human–machine interaction, etc. The various factors that need to be considered for gesture recognition are speech, noise and object position.

Image captioning

Image captioning is one upcoming research area of computing that considers the generation of captions based on images [21]. Image captioning identifies the objects in the given image it adopts supervised learning methods wherein the set of images and its captions are first put in the training set and then tested with other images for checking the image captions. It has been demonstrated in various applications such as recommendation, image indexing and image annotation. Some of the works has already been explored in this field.

Video question answering

Video question answering (VQA) is one of the promising research fields where the answer based on a question is given using the temporal/spatial information in the video [22]. A correlation analysis between the spatial and temporal features in the video is used for answering. It is divided into three subtasks, namely endpoint identification, correlation analysis and reasoning. The reasoning schemes are dependent on the audio sequences that are present in the video. There are some existing works that focus on the aspect of VQA that are discussed as follows.

1.5 Challenges of Multi-Modal Analytics

Multi-modal analytics requires data from different modalities to be analyzed, and a fusion strategy is employed for the final classification. The various applications of multi-modal analytics including healthcare, gesture recognition, image captioning and video question answering were discussed in the previous section. The overhead of inclusion of various modalities is a common challenge in multi-modal analytics [23]. The various challenges that are faced with multi-modal analytics are discussed as follows.

Dimensionality and data conflict

Multi-modal data exists in various sources and formats. The variation in the formats provides a challenge for extracting valuable information from the data. However, the large dimension of the data between the modalities also is an overhead for the analysis, computation and memory consumption. The synchronization between the formats and the dimensions of data needs to be considered for the multi-modal analysis. As discussed in the previous sections, feature-level fusion is more flexible than decision-level fusion. Hence, some of the dimensionality reduction methods like PCA and k-NN can be used for improved analysis.

Data availability

The most significant challenge of multi-modal analytics is data availability. The different types of multi-modalities of image, audio and video might not be available at a significant time. In some cases, there can be more training samples in audio data than the video. Hence, it affects the overall performance of the multi-modal analytics as the networks trained with limited datasets will not scale well. However, there are sampling methods and data augmentation methods that can be used to increase the size of the dataset.

Real-time processing and scalability

The performance and scalability of multi-modal analytics applications depend on real-time data and processing as well. The complex networks used for the multi-modal data might lead to a trade-off between accuracy and efficiency. Hence, the deterioration of the accuracy in the networks might be because of the reduced computing capacity. Applications based on computer vision, object detection and other applications require more computing power. However, the challenge can be mitigated by the use of cloud/edge solutions in such applications.

References

1. Pouyanfar, S., Yang, Y., Chen, S. C., Shyu, M. L., & Iyengar, S. S. (2018). Multimedia big data analytics: A survey. *ACM Computing Surveys (CSUR)*, 51(1), 1–34.
2. Fong, B., & Westerink, J. (2012). Affective computing in consumer electronics. *IEEE Transactions on Affective Computing*, 3(2), 129–131.

3. Abbasi, A., Chen, H., & Salem, A. (2008). Sentiment analysis in multiple languages: Feature selection for opinion classification in web forums. *ACM Transactions on Information Systems (TOIS)*, 26(3), 1–34.
4. Atrey, P. K., Hossain, M. A., El Saddik, A., & Kankanhalli, M. S. (2010). Multimodal fusion for multimedia analysis: A survey. *Multimedia Systems*, 16(6), 345–379.
5. Wang, Y., Liu, Z., & Huang, J. C. (2000). Multimedia content analysis-using both audio and visual clues. *IEEE Signal Processing Magazine*, 17(6), 12–36.
6. Poria, S., Cambria, E., Bajpai, R., & Hussain, A. (2017). A review of affective computing: From unimodal analysis to multimodal fusion. *Information Fusion*, 37, 98–125.
7. Wang, J., Kankanhalli, M. S., Yan, W. Q., & Jain, R. (2003). Experiential sampling for video surveillance. In *ACM Workshop on Video Surveillance*. Berkeley.
8. Adam, E., Mutanga, O., Odindi, J., & Abdel-Rahman, E. M. (2014). Land-use/cover classification in a heterogeneous coastal landscape using RapidEye imagery: Evaluating the performance of random forest and support vector machines classifiers. *International Journal of Remote Sensing*, 35(10), 3440–3458.
9. Luo, R. C., Yih, C. C., & Su, K. L. (2002). Multisensor fusion and integration: Approaches, applications, and future research directions. *IEEE Sensors Journal*, 2(2), 107–119.
10. Goh, G., Cammarata, N., Voss, C., Carter, S., Petrov, M., Schubert, L., & Olah, C. (2021). Multimodal neurons in artificial neural networks. *Distill*, 6(3), e30.
11. Bayoudh, K., Knani, R., Hamdaoui, F., & Mtibaa, A. (2021). A survey on deep multimodal learning for computer vision: Advances, trends, applications, and datasets. *The Visual Computer*, 38, 1–32.
12. O'Halloran, K. L., Tan, S., Pham, D. S., Bateman, J., & Vande Moere, A. (2018). A digital mixed methods research design: Integrating multimodal analysis with data mining and information visualization for big data analytics. *Journal of Mixed Methods Research*, 12(1), 11–30.
13. Russakovsky, O., Deng, J., Su, H., Krause, J., Satheesh, S., Ma, S., … & Fei-Fei, L. (2015). Imagenet large scale visual recognition challenge. *International Journal of Computer Vision*, 115(3), 211–252.
14. He, K., Zhang, X., Ren, S., & Sun, J. (2016, October). Identity mappings in deep residual networks. In *EurAUopean conference on computer vision* (pp. 630–645). Springer, Cham.
15. Elharrouss, O., Almaadeed, N., & Al-Maadeed, S. (2021). A review of video surveillance systems. *Journal of Visual Communication and Image Representation*, 77, 103116.
16. Yeong, D. J., Velasco-Hernandez, G., Barry, J., & Walsh, J. (2021). Sensor and sensor fusion technology in autonomous vehicles: A review. *Sensors*, 21(6), 2140.
17. Abbaschian, B. J., Sierra-Sosa, D., & Elmaghraby, A. (2021). Deep learning techniques for speech emotion recognition, from databases to models. *Sensors*, 21(4), 1249.
18. Arul, R., Al-Otaibi, Y. D., Alnumay, W. S., Tariq, U., Shoaib, U., & Piran, M. D. (2021). Multi-modal secure healthcare data dissemination framework using blockchain in IoMT. *Personal and Ubiquitous Computing*, 1–13. https://doi.org/10.1007/s00779-021-01527-2

19. Chen, X., Xu, S., Ji, Q., & Cao, S. (2021). A dataset and benchmark towards multi-modal face anti-spoofing under surveillance scenarios. *IEEE Access, 9,* 28140–28155.

20. Duan, H., Sun, Y., Cheng, W., Jiang, D., Yun, J., Liu, Y., & Zhou, D. (2021). Gesture recognition based on multi-modal feature weight. *Concurrency and Computation: Practice and Experience, 33*(5), e5991.

21. Ji, J., Luo, Y., Sun, X., Chen, F., Luo, G., Wu, Y., & Ji, R. (2021, May). Improving image captioning by leveraging intra-and inter-layer global representation in transformer network. *Proceedings of the AAAI Conference on Artificial Intelligence* (Vol. 35, No. 2, pp. 1655–1663).

22. Wu, Y., Ma, Y., & Wan, S. (2021). Multi-scale relation reasoning for multi-modal visual question answering. *Signal Processing: Image Communication, 96,* 116319.

23. Wang, Y. (2021). Survey on deep multi-modal data analytics: Collaboration, rivalry, and fusion. *ACM Transactions on Multimedia Computing, Communications, and Applications (TOMM), 17*(1s), 1–25.

Exercises

1. With a neat diagram, discuss the various features that need attention for the feature-level fusion.

2. Discuss decision-level fusion with a neat diagram.

3. Explain in brief the different types of applications based on text analytics.

4. Discuss the cloud applications of multi-modal analytics.

5. Describe the challenges of multi-modal analytics.

2

Cloud and Deep Learning

2.1 Overview of Cloud

Cloud computing is a paradigm that offers services in terms of storage, processing and networks. It provides the pay-as-you-go model based on the consumption of various services. The key idea of the cloud is to make available the computational power for large-scale consumption in various applications [1]. This makes it a viable option for scaling operations as the cost per unit is minimized. There are various companies that offer cloud services such as Google, IBM, Oracle and Microsoft. Each of them offers in various capabilities the storage, processing and networking in their own platforms. However, the principles of offerings remain the same with additional capabilities of databases, data analytics and artificial intelligence (AI) solutions. We discuss the various categories of cloud solutions, neural network foundations and other topics in this chapter. The focus of this chapter is to introduce the Google Cloud solutions available for neural networks and deep learning.

Cloud solutions are required for various applications such as education, e-commerce, retail, logistics, energy and e-government services. The solutions preferred vary from time to time and application wise. However, there are three solutions [2] that every application scenario looks up to the cloud. The general categories of cloud solutions are listed as follows as shown in Figure 2.1.

- **Public cloud:** It is one of the conventional computing models offered in the cloud where services of computing and infrastructure are offered publicly to enterprises and individuals. The cloud service provider is responsible for the management of configuration and services. The services are limited in this solution as the public services will not ensure privacy.

- **Private cloud:** It is a model offered with services of computing and infrastructure where an organization is responsible for the management of configuration and services. Usually, the services are provisioned within the campus of the organization or through a private

DOI: 10.1201/9781003215974-3

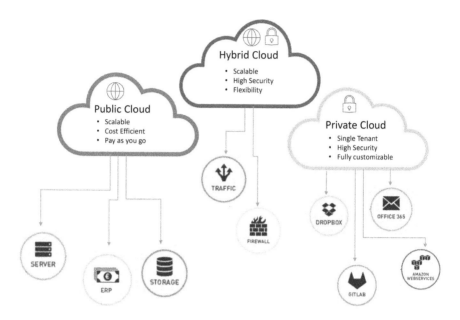

FIGURE 2.1
Cloud categories.

network by the cloud service provider. The cost of the services is high because of the private network.

- **Hybrid cloud:** It is a model of services between the cost efficiency of public cloud and data integrity of the private cloud. It provides a multi-cloud strategy wherein the porting and sharing of data take place on premises of the campus and the application is public.

Once, the solution offering is finalized based on the application scenario, the services offered need to be ensured. For example, infrastructure alone can be the service offered in an application that requires virtual machines (VM) instances. In this regard, the various service delivery methods offered within the cloud solutions are listed as follows as shown in Figure 2.2.

- **Infrastructure as a Service (IaaS):** It is a service offered by the cloud service provider to host the data and applications using the required infrastructure such as virtual machines. A necessary system administration is required to manage the infrastructure to maintain the data and applications.
- **Platform as a Service (PaaS):** It is a service offered by the cloud service provider, where the required system and development tools are managed with the hardware. The user/client can focus on the

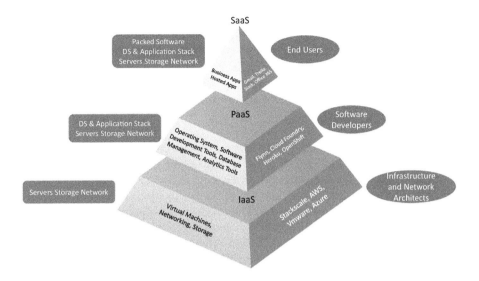

FIGURE 2.2
Cloud delivery models.

business logic and the workflow that forms the essential needs of an application.

- **Software as a Service (SaaS):** It is the most recognizable service that is used widely in the management of computing, infrastructure and services. Examples include Gmail, Outlook, Google Drive and Google Photos.

2.1.1 Google Cloud and Services

There are various services that are offered with the Google Cloud such as storage, services and analysis of data [3]. The ecosystem of the Google Cloud is designed to perform different operations within a secure perimeter. It includes different services like compute, storage, big data/analytics, AI and other networking, developer, and management services. The services offered by Google Cloud are briefly explained as follows.

Cloud Compute
The compute service in the Google Cloud offers different types of products based on the computational needs of the application. The products consist of Compute Engine, App Engine, Kubernetes Engine, Container Registry, Serverless Cloud and Cloud Run. Compute Engine offers different virtual

instances of computing, as shown in Figure 2.3. App Engine provides a platform for the development of applications for mobile, IoT and web. Kubernetes Engine provides an orchestration manager for docker containers and customization. Container Registry provides private storage using Serverless Cloud functions for connection. Cloud Run offers a service for auto-scaling of the containers that are used for applications in the cloud.

Cloud Storage
Cloud Storage provides storage service that is scalable, gives real-time access and allows archival of data. It is different from Google Drive. The Cloud Storage service provides consistent, secure access within the cloud perimeter according to the demand of the applications. Cloud Storage services are depicted in Figure 2.4. The products available in the cloud storage are Cloud Storage, Cloud SQL, Cloud BigTable, Cloud Spanner, Cloud Datastore and Persistent Disk. Cloud Storage is the general-purpose platform offered within the cloud for different applications. Cloud SQL manages the MySQL and PostgreSQL queries within the applications. Cloud

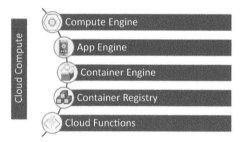

FIGURE 2.3
Cloud Compute services.

FIGURE 2.4
Cloud Storage services.

BigTable offers NoSQL services with petabyte-sized storage and access for different applications. Cloud Datastore offers transaction-oriented NoSQL database using Persistent Disk solution.

Big Data and Analytics
Google Cloud offers various services of big data and analytics for data exploration, data warehousing, Hadoop solutions and messaging systems. The different types of big data services offered by the Google Cloud are Cloud BigQuery, Cloud Dataproc, Cloud Dataflow, Cloud Dataprep, Cloud Datastudio, Cloud Datalab and Cloud Pub/Sub. Big data and analytics in Google Cloud service offerings are shown in Figure 2.5. Cloud BigQuery is a service offered for serverless data warehousing. Cloud Dataproc offers a service of managed Hadoop/Apache Spark infrastructure for managing distributed applications. Cloud Dataflow offers a service for data transformation for batch processing of Hadoop/Spark applications. Cloud Dataprep offers a service for serverless infrastructure for data analytics. Cloud Datalab provides a notebook environment for machine learning connecting Cloud Datastudio for visualization and report dashboards. Cloud Pub/Sub offers a messageless service infrastructure.

Cloud Artificial Intelligence (AI)
Google Cloud AI provides different services for training models for custom AI tasks with different Rest APIs. The different types of products in the Cloud AI are Cloud AutoML, Cloud Machine Learning Engine, Cloud TPU, Cloud Natural Language API, Cloud Speech API, Cloud Vision API, Cloud Translate API and Cloud Video Intelligence API. Figure 2.6 shows AI in Google Cloud. Cloud AutoML provides services for the transfer of learning by leveraging machine learning models. Cloud Machine Learning Engine provides the service for the deployment of machine learning

FIGURE 2.5
Big data and analytics in Google Cloud.

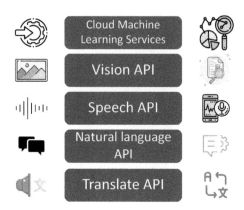

FIGURE 2.6
AI in Google Cloud.

models and distributed training. Video Intelligence combined with Cloud TPU provides the service for large-scale training of videos. Cloud Vision API offers the service for classification and segmentation of images with cloud translate API for the translation of languages.

Google Cloud Setup and Configuration
A command-line interface (CLI) is provided by the Google Cloud Platform (GCP) for interacting with various services offered. The resources of the GCP can also be accessed with web-based CLI using Google Cloud software development kit (SDK). The GCP SDK needs to be installed on the local machine to interact with the GCP. The SDK also provides developers to install the libraries and work with various products and services. There are various shell commands included in the SDK with tools. Some of the major tools are listed as follows.

- **Gcloud tool:** It manages interactions with the GCP, including authentication and configuration.
- **Gsutil tool:** It manages the responsibility of the interaction of Cloud Storage buckets and objects.
- **Bq tool:** It is used for interaction and management of Google Big Query.
- **Kubectl tool:** It is used for management of Kubernetes containers in the GCP cluster.

The scope of this book is to use the GCP for deep learning and multi-modal analytics. The next sections of the book explain the required components of GCP with set-up and configuration for multi-modal analytics.

Setting Up an Account on Google Cloud Platform

In this section, the account set-up and configuration in GCP is discussed. A GCP account gives access to the platform and its service with $300 credit that is valid for 12 months. Figure 2.7 depicts dashboard of GCP, and the steps for setting up the account are listed as follows:

1. Open an account using the URL https://cloud.google.com/

2. The necessary details of identity, address and card need to be filled.

3. The account creation takes a moment; wait for it.

4. Welcome to GCP page is presented once the account is created.

5. Cloud dashboard can be seen using the option 'Home' present on the left corner. The dashboard gives a bird's eye view of projects and its summary.

Google Cloud SDK

The SDK is needed to interact with the GCP and its various services. The necessary steps required for configuration of the Cloud SDK are listed as follows.

1. Download and install the appropriate Cloud SDK using the URL https://cloud.google.com/sdk/, Figure 2.8 shows the same.

2. Install the Google Cloud SDK and its default components with necessary OS instructions. Figure 2.9 shows the step of OS instructions in GCP SDK.

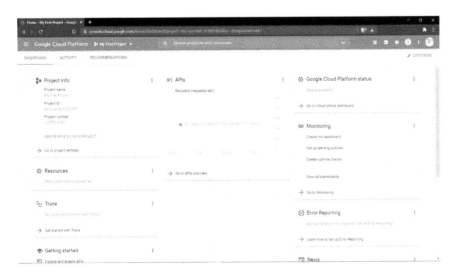

FIGURE 2.7
GCP dashboard.

Cloud-based Multi-Modal Data and Analytics

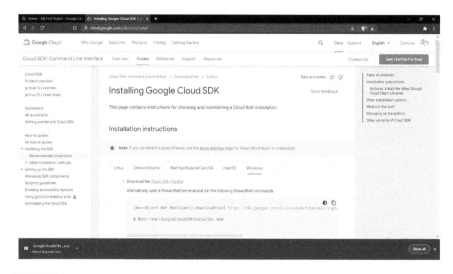

FIGURE 2.8
GCP SDK.

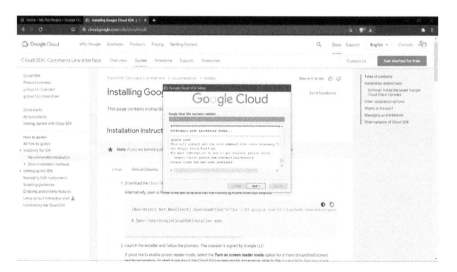

FIGURE 2.9
GCP SDK – OS instructions.

3. Cloud SDK can be initiated using the terminal with the command 'gcloud init'.

4. Open the terminal application of your OS and run the command 'gcloud init' to begin authorization and configuration of the Cloud SDK. The SDK is successfully installed with the message of successful authentication.

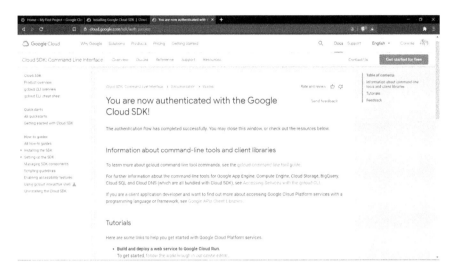

FIGURE 2.10
GCP authentication.

5. Once, the authentication is successful, the required project can be selected for usage. Figure 2.10 displays the authentication phase.

2.1.2 Google Cloud Storage (GCS)

GCS is a product in the Google Cloud used for the storage of data for diverse applications. It can be used to store both live and archived data. It provides consistency, durability, scalability and high availability. It stores the data at multiple locations to prevent the loss of data, but the query returns the most updated version of the data. The options and features available within the GCS for storage are listed as follows.

- Memcache can be used through the App Engine.
- The static objects of size 5GB are supported by the GCS.
- Cloud Spanner can be used in conjunction with database applications.
- Low Latency memory is provided with the GCS for development of persistent applications.
- It supports the data for CloudSQL, BigTable and NoSQL databases.

GCS and Bucket
Bucket is the basic structure used for Cloud Storage in the GCP. Buckets can be organized hierarchically based on the applications for storing the data. The steps for creating a bucket in GCP are listed as follows.

1. The option 'Create Bucket' need to be selected in the Cloud dashboard. Figure 2.11 shows the GCS Cloud dashboard.

2. A unique name needs to be given to the bucket. The unique name is given to ensure that no two buckets are of the same name.

3. The next step is to select the storage class for the created bucket. There are two options namely multi-region class and cold-line storage class. Cold-line class is usually used for the backup files, whereas the multi-region class is used to access the data all over the world.

4. The bucket creation is completed using the option 'Create' in the GCS; this step is shown in Figure 2.12.

Upload, Delete Data to a Bucket in GCS

GCS supports uploading of individual files/folders into the available buckets. The following steps show the uploading of a local file into a bucket in GCS, Figures 2.13–2.16 depict the process of uploading files into available buckets:

1. Select the bucket to which the files/folders need to be uploaded.

2. Select the option 'Upload Files'.

3. Select the required file from the local machine and upload.

4. The upload can be verified in the bucket.

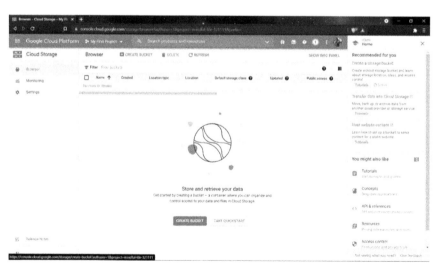

FIGURE 2.11
GCS – Cloud dashboard.

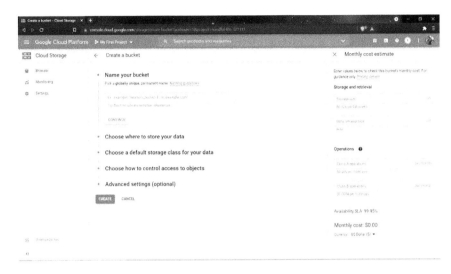

FIGURE 2.12
GCS – Creation of bucket.

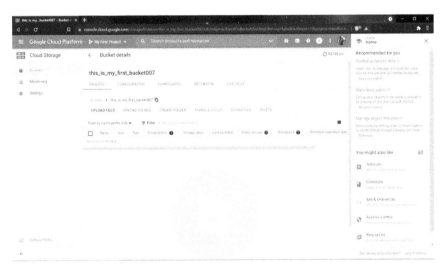

FIGURE 2.13
GCS – Selecting files to upload.

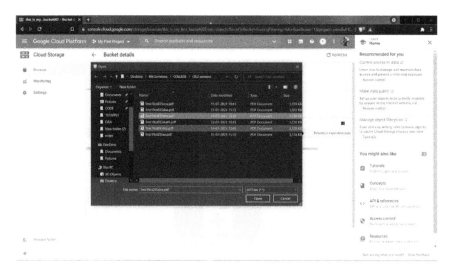

FIGURE 2.14
GCS – Select required file from local machine.

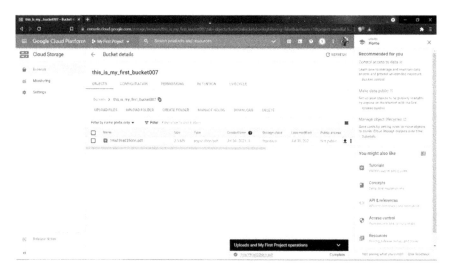

FIGURE 2.15
GCS – Uploaded file in bucket.

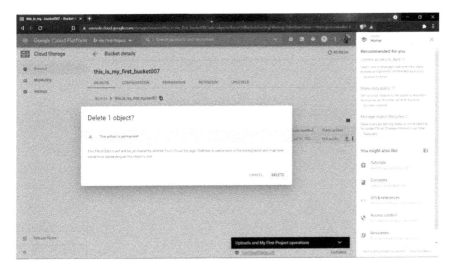

FIGURE 2.16
GCS – Deleting files from bucket.

1. The file that needs to be deleted can be checked to delete it from the bucket.

Managing GCS with CLI

The GCS can be managed and used with CLI. The options and the commands used are listed as follows.

- **Bucket creation:** The command used for the creation of the bucket is 'mb'. It expects the argument as the bucket name.

```
gsutil mb gs://<bucket_name>
```

- **Listing buckets:** The buckets that are available in the GCP account can be seen using the command 'ls'. A small example is shown in Figure 2.17.

- **Upload data:** The 'cp' command is used to upload the data into the buckets in GCP. It expects the arguments source file destination and the bucket destination.

```
gsutil cp -r <source_file_location>
gs://<destination_bucket>
```

- **Delete data:** The 'rm' command is used to delete the data in the bucket in GCP.

```
gsutil rm -r gs://<destination_bucket>/<file_name>
```

FIGURE 2.17
GCS – gsutil ls.

2.1.3 Google Compute Engine (GCE)

GCE provides the configuration and management of VMs in the GCP. These VMs run on the Google data centers and take the advantage of the advanced processing capabilities. They are able to scale up and increase their performance along with load balancing of various applications. The users are provided the option of new VMs or out-of-box VMs based on the needs of the applications/users. The user in the GCP pays only for the time the VM is used. In this section, the configuration of the VM and its management are discussed.

Creating VM in GCP

The following steps are involved in VM creation in GCP. Figures 2.18–2.21 show the same:

1. Select the option 'VM instances' in the 'Compute Engine' of the GCP dashboard.
2. Select the instance name, machine type, region and the zone in which it should be up and running.
3. Select the OS image that is needed for the VM instance created. It provides various options with Ubuntu, Suse, etc.
4. Select 'Allow HTTP traffic' for intercommunications. The VM instance is successfully created as shown in the figure once the create button is selected.

FIGURE 2.18
GCE – Dashboard.

FIGURE 2.19
GCE – Select instance to run.

FIGURE 2.20
GCE – Selecting OS for VM instance.

FIGURE 2.21
GCE – VM instance created.

VM connection and deletion in GCP:

Steps involved in VM connection and deletion in GCP are explained below with the help of Figures 2.22 and 2.23:

1. In the GCP dashboard, select the VM that is needed and select the option 'SSH'. It launches the terminal to access the VM.

FIGURE 2.22
GCP – SSH.

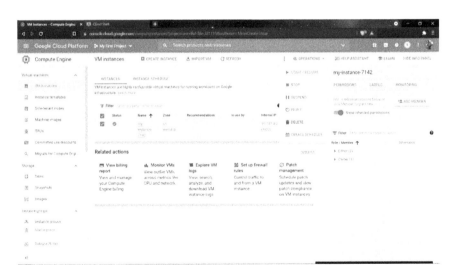

FIGURE 2.23
GCP – VM deletion.

2. The VM instance can be deleted if it is no longer needed using the option 'delete' next to the VM instance.

2.1.4 Google Colaboratory

It is often referred to as 'Google Colab' or simply 'Colab' for developing prototypes of machine learning/deep learning models. It provides powerful access to hardware options with GPUs and TPUs. It provides a serverless environment for the developers to prototype the applications based on deep learning. The steps to use the Colab are listed as follows with the help of Figures 2.24–2.26:

Colab configuration

1. With the existing google account, visit the URL https://colab. research.google.com/. It displays the welcome page as shown in the figure.
2. Select the 'New Notebook' option and type 'print('Hello')' to run the notebook as shown in the figure.
3. The runtime option of CPU/GPU can be selected using the runtime settings, as shown in the figure.
4. The notebooks created on the Colab can be saved locally/github/ drive with various options, as shown in the figure.

FIGURE 2.24
Colab configuration.

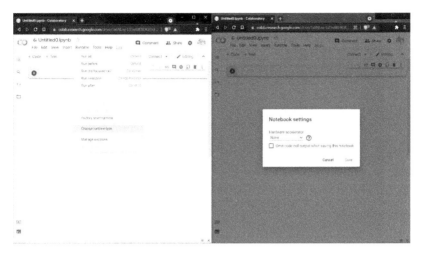

FIGURE 2.25
Colab runtime settings.

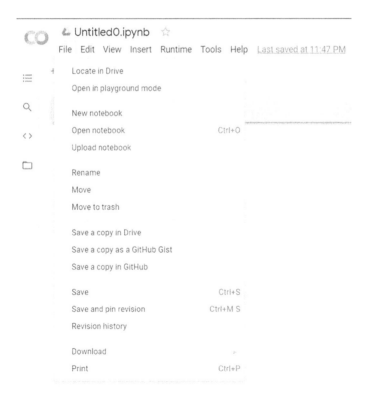

FIGURE 2.26
Saving of Colab.

References

1. Rambabu, M., Gupta, S., & Singh, R. S. (2021). Data mining in cloud computing: survey. In *Innovations in Computational Intelligence and Computer Vision* (pp. 48–56). Springer, Singapore.
2. Atieh, A. T. (2021). The next generation cloud technologies: A review on distributed cloud, fog and edge computing and their opportunities and challenges. *ResearchBerg Review of Science and Technology*, 1(1), 1–15.
3. https://cloud.google.com/

Exercises

1. Describe the three general categories of cloud solutions
2. Describe the three service delivery methods offered within cloud solutions.
3. List the steps to set up an account on the Google Cloud Platform.
4. What is Google Cloud Storage (GCS)? What are the options and features available within the GCS for storage?
5. Discuss four commands of the CLI that can be used to manage GCS.

3

Overview of Deep Learning

Deep learning applications are based on neural networks for learning and prediction for various problems of computer vision, self-driving, speech translation and other applications. Neural networks are the mathematical models inspired by the human brain that facilitate learning and prediction [1]. Human brain has the ability to learn various tasks that are non-trivial. The complex learning process and the ability to acquire it set us apart as intelligent beings. Some examples of the tasks include face recognition in a millionth of second, linguistic representations and adaptability to learn different languages at a time. The challenge of the current artificial intelligence (AI) systems is to understand the structural patterns behind this and learn efficient neural networks. Humans are often inspired by nature and the things around it like bird to airplane. Likewise, human brain can be viewed as a society with many agents that are responsible for many activities. These agents are the neurons that are of interest for learning and representations. In this section, the different parts of neuron and its relation to deep learning is explained.

3.1 Neural Networks and Their Foundations

Neural networks play an important role in deep learning and its various applications [2]. Large deep networks consist of neural networks as the core component for the tasks of classification and others. Artificial neural network (ANN) forms the basis for various neural networks. It provides a representation and the architecture for deep learning applications. An ANN is composed of three important components, as shown in Figure 3.1.

- **Input layer**: It is the layer that collects the data from single or different sources. The initial pre-processing steps are done before feeding the data to the input layer. The usual pattern of feeding the data to the input layer is in the form of training and testing. The training data is learned from different computations by ANN, and testing data is used for prediction purpose.

DOI: 10.1201/9781003215974-4

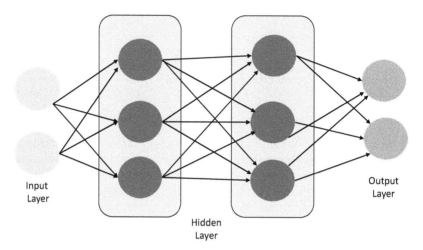

FIGURE 3.1
Artificial neural network and components.

- **Hidden layer**: It is known as the workhorse layer of deep learning models. It consists of multiple neurons and layers for learning. The width of the hidden layer network and the topology depends on the application and the data. Each layer gives a sophisticated set of feature representations for the input data considered. The feature representations are then given to the output layer for the final prediction.

- **Output layer**: It is the last layer of the ANN where the prediction of the label or class will happen. The different types of functions like Softmax are employed here for the prediction of the class.

The layers in the ANN are dependent on the weights and the interconnections between them. The learning representations of the data need to be assigned weights and activation functions, which are responsible for firing the required values to the output layer. The different types of activation functions and weights are discussed in the next sections.

3.1.1 Weights and Activation Function

Every neuron in the ANN is assigned a weight for learning. Initially, the weights are assigned randomly and then updated for better learning. The general procedure of learning with weights is shown in Figure 3.2. The neuron weights in the input layer are multiplied with the inputs and passed to the activation function. The output of the activation function is then passed to the next layer of the neural network. Summation and activation units are

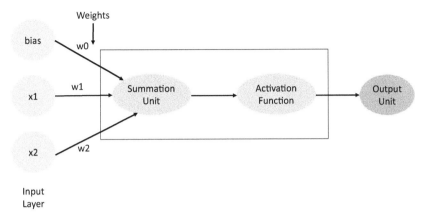

FIGURE 3.2
Weights and activation function in ANN.

shown in Equations (3.1) and (3.2). The next layer can be a hidden layer/output layer. Every neuron is also assigned a bias (typically set to 1) for controlling the weighted sum of the activation. The different types of activation functions and the working of an ANN are explained in further sections.

$$\text{Summation Unit} = \sum w_0\, b + w_1\, x_1 + w_2\, x_2 \tag{3.1}$$

$$\text{Activation Unit} = \varnothing(\text{Summation Unit}) \tag{3.2}$$

Activation functions resemble the communication between the neurons, that is, responding to the stimuli based on the threshold [3]. The activation function in the ANN checks if the neuron has the correct result for the prediction in the output layer. The different types of activation functions are discussed as follows.

- **Sigmoid**: It is a non-linear function where the activations fall in the range of 0 and 1. The output of the activation is fired when the threshold is above 0.5. Figure 3.3 describes the sigmoid function. The drawback of the sigmoid activation function is the concentration of values either at 0 or 1, which results in vanishing gradient problem. The activation function is represented using Equation (3.3). The code snippet shows the visualization of sigmoid function values that fall in the range of 0 and 1.

$$f(x) = \frac{1}{1 + e^{-x}} \tag{3.3}$$

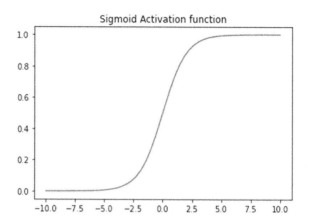

FIGURE 3.3
Sigmoid activation function.

```
import numpy as np
import matplotlib.pyplot as plt
import numpy as np

def sigmoid(v):
    return 1/(1+np.exp(-v))

v = np.linspace(-10, 10)
plt.plot(v, sigmoid(v))
plt.axis('tight')
plt.title('Sigmoid Activation function')
plt.show()
```

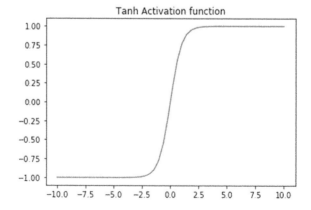

FIGURE 3.4
Tanh activation function.

- **Hyperbolic Tangent (tanh):** It is a function with the range of values from −1 to 1. Figure 3.4 explains the tanh activation function. The outputs of the function are zero-centered, as shown in Equation (3.4). The code snippet shows the tanh function with the range of values from −1 to 1.

$$f(x) = \frac{2}{1+e^{-2x}} - 1 \tag{3.4}$$

```
import numpy as np
import matplotlib.pyplot as plt
import numpy as np

def tanh(x):
    return (np.exp(x)-np.exp(-x))/(np.exp(x)+np.
        exp(-x))

x = np.linspace(-10, 10)
plt.plot(x, tanh(x))
plt.axis('tight')
plt.title('Tanh Activation function')
plt.show()
```

- **Rectified linear unit (ReLU):** It is a function wherein the value is 0 for $x < 0$, and a linear slope for the values of $x > 0$ is shown in Equation (3.5). The code snippet shows the ReLU function, and Figure 3.5 shows the Relu activation function.

$$f(x) = \max(0, x) \tag{3.5}$$

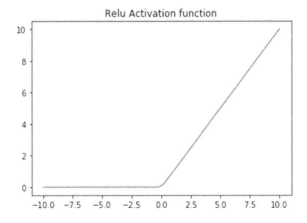

FIGURE 3.5
Relu activation function.

```
import numpy as np
import matplotlib.pyplot as plt
import numpy as np

def relu(y):
    out = []
    for i in y:
        out.append(max(0,i))
    return out

y = np.linspace(-10, 10)
plt.plot(y, relu(y))
plt.axis('tight')
plt.title('Relu Activation function')
plt.show()
```

3.1.2 Loss Function for Training

The main goal of any neural network is to minimize the error of learning. The two main functions used for measuring the error are mean squared error (MSE) and cross-entropy loss function. MSE estimates the sum of squared difference between the actual and predicted value of the problem considered, as shown in Equation (3.6). Meanwhile, cross-entropy function estimates the difference between the actual and predicted probability values for the problem considered. An example of estimation using MSE function is as shown in Figure 3.6. Here, the actual output is 5.3 and the predicted output is 5.0 with the inputs considered 5.4 and 5. The MSE is calculated using the equation, where 'n' is the number of neurons in the output layer. The value of the MSE is used for the assessment of the network's output.

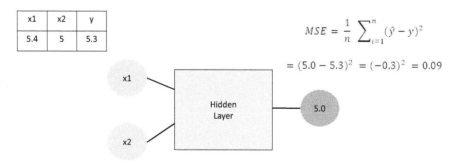

x1	x2	y
5.4	5	5.3

$$MSE = \frac{1}{n}\sum_{i=1}^{n}(\hat{y}-y)^2$$

$$= (5.0-5.3)^2 = (-0.3)^2 = 0.09$$

FIGURE 3.6
Training using loss function in ANN.

$$\text{MSE} = \frac{1}{n} \sum_{i=1}^{n} (\hat{y} - y)^2 \tag{3.6}$$

$$= (5.0 - 5.3)^2 = (-0.3)^2 = 0.09$$

3.1.3 Training a Neural Network

A neural network is trained with the basis of back propagation. The basic steps involved in training of the neural network are explained in the following sections.

- Network Initialization
- Forward Propagation
- Back Propagation
- Training Neural Network
- Prediction

3.1.3.1 Network Initialization

A neural network essentially comprises a combination of an input layer, hidden layer and output layer in that order. Each neuron in the neural network needs to be initialized with the weights before training. For each input connection, one additional weight bias is considered. An example of neural network considered is shown in Figure 3.7. It consists of one input layer (2 neurons), one hidden layer (1 neuron) and one output layer (2 neurons).

In the section of the code considered, a dictionary 'weights' is used to store the properties of neurons in each layer. The dictionaries are arranged in the network of layers for further training. The weights are initialized with random

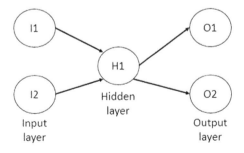

FIGURE 3.7
Neural network.

numbers from 0 to 1. The function 'initialize_network()' considers three parameters namely input s, number of hidden layer neurons and number of output s. An additional weight bias is considered with I as the input weight.

```
from random import seed
import numpy as np

def init_neural _net (input _neurons , hidden_neurons ,
    output _neurons) :
m ral_net = lis t()
hidden_layer = [{' weights' : [np.random.rand() for
n in range( input_neurons + 1 ) ]} for n in range(
hidden_neurons) ]
neural_ net.append(hidden_layer)

output_layer = [{'weights ': [np.random.rand()
for n in range(hidden_neurons + 1 )]} for n in
range(output_neurons)]
neural_net.append(output_layer) return neural_net
np .ra ndo m.seed( 1 )
neural_net = init_neural_net(2, 1, 2)
for layer in neural_ne t:
    print( layer)
```

Output

[{'weights': [0.417022004702574, 0.7203244934421581, 0.00011437481734488664]}]

[{'weights': [0.30233257263183977, 0.14675589081711304]}, {'weights': [0.0923385947687978, 0.1862602113776709]}]

3.1.3.2 Forward Propagation

Once the weights are initialized in the neural network, the next step is to prepare the neurons for activation and firing through forward propagation. The propagation is divided into three parts, namely, activation, transfer and propagation. The activation of the neuron is calculated as a sum of the weights using Equation (3.7). The 'weights' considered previously has the bias weight as the last element. The code snippet for the activation is shown below. It considers the bias weight first and then the summation of all the weights in each layer, as shown in the equation.

$$A = \sum_{i \in I} w_i I_i + b \qquad (3.7)$$

```
def apply_activation(weights,inputs):
    output = weights[-1]
    output += sum([weights[il * inputs[il for i    in
        range(len(weights)-1])
    return output
```

The summation of the weights is followed by the transfer functions. For a neural network, the different transfer functions like sigmoid function, hyperbolic tangent function and rectifier transfer function are used. The code snippet of the 'transfer' function is as shown below. The transfer function will take the activation value and produce a value of the respective transfer function. The transfer function considered here is the sigmoid activation function, as shown in Equation (3.8).

$$T = \frac{1}{1+e^{-A}} \tag{3.8}$$

```
def apply_sigmoid(activation):
    return 1 .0 / (1 .0 + np .exp( -activation))
```

The final step of the forward propagation is to use the activation and transfer functions as defined above to produce the final value of the outputs. The code snippet for the forward propagation is as shown below with the function forward_propagate(). The function considers each layer, and each neuron in the layers with activate and transfer functions.

The function is run with the sample input pattern [1,0] and produces the output of the values that are not closer than the actual ones. It shows that the random weights considered need to be updated for better training and learning. The training of the weights is done using the back propagation, which is discussed in the next section.

```
def apply_activation(weights,inputs):
    output = weights[-1]
    output += sum([weights[il * inputs[il for i    in
        range(len(weights)-1])
    return output

def apply_sigmoid(activation):
    return 1 .0 / (1 .0 + np .exp( -activation))

def forward_propagation (neural_net, input_sample):
    propagated_output = input_sample
    for neural_layer in neural _net:
        new_layer_inputs = []
        for neuron_cell in neural_layer:
```

```
            activation_output = apply_
                activation(neuron_cell ['weights'],
                propagated_output)
            neuron_cell [ 'output '] = apply_
                sigmoid(activation_output )
            new_layer_inputs.append(neuron_cell
                ['output'])
        propagated_output = new_layer_inputs
    return propagated_output
```

Output

neural_net = [[{'weights': [0.13436424411240122, 0.8474337369372327, 0.763774618976614]}],

[{'weights': [0.2550690257394217, 0.49543508709194095]}, {'weights': [0.4494910647887381, 0.651592972722763]}]]]

row = [1, 0, None]

output = forward_propagation(neural_net, row)

print(output)

[0.6629970129852887, 0.7253160725279748]

3.1.3.3 Backward Propagation

In this section, the back-propagation steps of updating the weights are explained. Initially, the error is computed between the desired outputs and the actual outputs of the forward propagation. The errors are then back-propagated through the model's neural network for updating the weights of the network.

3.1.3.3.1 Transfer Derivative

The activation function used is sigmoid. First, the derivative of the sigmoid function is obtained for further back propagation using Equation (3.9).

$$\sigma'(x) = \sigma(x)(1 - \sigma(x)) \qquad (3.9)$$

```
  def sigmoid_derivation_function(x):
  return x* (1.0 - x)
```

3.1.3.3.2 Error Back Propagation

In this section, the back-propagation steps are discussed with necessary equations (3.10, 3.11, 3.12, 3.13, 3.14). The first step of the back-propagation is the calculation of error in the output layers. In the considered example, there are two neurons in the output layer. For each of the output neuron, the error is calculated between the actual output and predicted output as shown in the equations. The activation function considered for the example is sigmoid activation function. The derivative of the activation function is obtained as shown in the equations. The error in the hidden node is calculated using the equation. The derivatives of the activation function of the corresponding hidden node and output node are obtained. Once the errors are calculated, the weight update is done afterward as discussed in the next section.

$$E_{o1} = \left(y_{o1} - \widehat{y_{o1}} \right) * \frac{\partial}{\partial T_1} \tag{3.10}$$

$$E_{o2} = \left(y_{o2} - \widehat{y_{o2}} \right) * \frac{\partial}{\partial T_2} \tag{3.11}$$

$$\frac{\partial}{\partial T_1} = O_1 * (1 - O_1) \tag{3.12}$$

$$\frac{\partial}{\partial T_2} = O_2 * (1 - O_2) \tag{3.13}$$

$$E_h = \nabla_{k \in H, \ j \in O} \left(w_k * e_j \right) * \frac{\partial}{\partial h_1} \tag{3.14}$$

The code snippet for the error calculation is as shown below. Each layer of the neural network is considered and errors are calculated using the derivative `sigmoid_derivation_function()`. The derivative is then used for the hidden layers. The procedure for the weight update is discussed in the next section.

```
def back_propagation_error(neural_net, desired):
    for layer_index in reversed(range(len
        (neural_net))):
        current_layer = neural_net(layer_index)
        errors = []
        if layer_index != len(neural_net) -1 :
            for neuron_index in range(len(current_
                layer)):
                current_layer_error = 0.0
                for back_prop_neuron in neural_net
                    [layer_index+1]:
```

```
                    current_layer_error+= (back_
                        prop_neuron['weights']
                        [neuron_index] *
                        back_prop_neuron['delta'])
                errors.append(current_layer_error)
        else:
            for neuron_index in range(len(current_
                layer)):
                neuron = current_layer[neuron_index]
                neuron['delta']=errors[neuron_
                    index]sigmoid_derivation_
                    function(neuron['output'])
        for neuron_index in range(len(current_
            layer)):
            neuron = current_layer['neuron_index']
            neuron['delta']=errors[neuron_
                index]* sigmoid_derivation_
                function(neuron['output'])
```

An example run of the above code with the below fixed neural network weights is shown below.

```
neural_net = [[{'output': 0.7105668883115941,
'weights': [0.13436424411240122, 0.8474337369372327,
0.763774618976614]}],
[{'output': 0.6213859615555266, 'weights':
[0.2550690257394217, 0.49543508709194095]},
{'output': 0.6573693455986976, 'weights':
[0.4494910647887381, 0.651592972722763]}]]
expected = [0, 1]
back_propagation_error(neural_net, expected)
for layer in neural_net:
    print(layer)
[{'output': 0.7105668883115941, 'weights':
[0.13436424411240122, 0.8474337369372327,
0.763774618976614], 'delta':
-0.0005348048046610517}]
[{'output': 0.6213859615555266, 'weights':
[0.2550690257394217, 0.49543508709194095],
'delta': -0.14619064683582808}, {'output':
0.6573693455986976, 'weights': [0.4494910647887381,
0.651592972722763], 'delta': 0.0771723774346327}]
```

3.1.3.3.3 Train Network

The network training is carried out with the steps forward propagation, backward propagation and weight updates. The weights of the network are updated using Equation (3.15), where 'ω' is the learning rate, δ is the derivative of neuron 'n' at 'j' layer and I is the input of neuron.

$$\Delta w_j(n) = \omega * \delta_j(n) * I_j(n) \tag{3.15}$$

The code snippet for the updating weights and training neural network is shown below. The weight update is carried out in the neural network using the function update_network_wieghts(). The line of code learning_rate * neuron_cell['delta'] * inputs[i] is used for updating the weights. The training of the network is carried out using the function `train_neural_net()`. The functions forward_propagation() back_propagation_error() update_network_weights() are used for training. In this way, the neural network can be trained. We show an example of random inputs for the analysis of training.

3.1.3.3.4 #Update Weights

```
def update_network_wieghts(neural_net, input_sample,
learning_rate):
    for layer_index in range(len(neural_net)):
        current_layer = neural_layer[layer_index]
        inputs = input_sample[:-1]

        if layer_index != 0 :
            inputs = [neuron_cell['output']
                for neuron_cell in
                neural_net[layer_index-1]]

        for nuron_cell in current_layer:
            for i in range(len(inputs)):
                neuron_cell['weights'][i] +=
                    learning_rate * neuron_
                    cell['delta'] * inputs[i]

            neuron_cell['weights'][-1] += learning_
                rate * neuron_cell['delta']
```

3.1.3.4 #Training Neural Network

```
def train_neural_net(neural_net, training_data,
    learning_rate, epochs, output_neurons):
    for epoch in range(epochs):
```

```
            total_error = 0
            for record in training_data:
                network_outputs = forward_
                    propagation(neural_net, record)
                desired = [0 for _ in range
                    (output_neurons)]
                desired[record[-1]] = 1
                total_error += sum([(desired[_] -
                    network_outputs[_]) ** 2 for _ in
                    range(len(desired))])

                back_propagation_error(neural_net,
                    desired)
                update_network_wieghts(neural_net,
                    record, learning_rate)
            print("Epoch : {0} learning_rate : {1}
                error : {2}".format(epoch, learning_rate,
                total_error))

np.random.seed(1)
dataset = [[2.7810836,2.550537003,0],
           [1.465489372,2.362125076,0],
           [3.396561688,4.400293529,0],
           [1.38807019,1.850220317,0],
           [3.06407232,3.005305973,0],
           [7.627531214,2.759262235,1],
           [5.332441248,2.088626775,1],
           [6.922596716,1.77106367,1],
           [8.675418651,-0.242068655,1],
           [7.673756466,3.508563011,1]]
input_neurons = len(dataset[0]) - 1
output_neurons = len(set([row[-1] for row in
    dataset]))
neural_net = init_neural_net(input_neurons, 2,
    output_neurons)
train_neural_net(neural_net, dataset, 0.5, 20,
    output_neurons)
for layer in neural_net:
    print(layer)

>epoch=0, lrate=0.500, error=6.189
>epoch=1, lrate=0.500, error=5.594
>epoch=2, lrate=0.500, error=5.397
>epoch=3, lrate=0.500, error=5.334
```

```
>epoch=4,  lrate=0.500,  error=5.303
>epoch=5,  lrate=0.500,  error=5.276
>epoch=6,  lrate=0.500,  error=5.244
>epoch=7,  lrate=0.500,  error=5.201
>epoch=8,  lrate=0.500,  error=5.140
>epoch=9,  lrate=0.500,  error=5.055
>epoch=10, lrate=0.500,  error=4.935
>epoch=11, lrate=0.500,  error=4.773
>epoch=12, lrate=0.500,  error=4.567
>epoch=13, lrate=0.500,  error=4.321
>epoch=14, lrate=0.500,  error=4.045
>epoch=15, lrate=0.500,  error=3.754
>epoch=16, lrate=0.500,  error=3.457
>epoch=17, lrate=0.500,  error=3.167
>epoch=18, lrate=0.500,  error=2.890
>epoch=19, lrate=0.500,  error=2.634


[{'weights': [0.31566864181142357,
0.6150057706819665, -0.038631187739985196],
'output': 0.9895049841068932, 'delta':
-0.00013254766839399567},

{'weights': [1.132849851892996, -1.4315991658357723,
-0.455674228702452], 'output': 0.9465461560938725,
'delta': 0.009437641300484592}]


[{'weights': [-0.0017292384329513286,
-1.6550808104847068, 0.5473448713504467], 'output':
0.2810319497792787, 'delta': -0.056783346576620916},

{'weights': [-0.17060500117194688,
1.685630528988771, -0.4047444898026321], 'output':
0.7191892189915582, 'delta': 0.056711446317097146}]
```

3.1.3.5 Prediction

This section explains the process of prediction based on the implementation of the neural network.

```
def predict-sample(neural_net, input_sample):
    network_outputs = forward_propagation(neural_
        net, input_sample)
    return network_outputs.
        index(max(network_outputs))
```

```
dataset = [[2.7810836,2.550537003,0],
           [1.465489372,2.362125076,0],
           [3.396561688,4.400293529,0],
           [1.38807019,1.850220317,0],
           [3.06407232,3.005305973,0],
           [7.627531214,2.759262235,1],
           [5.332441248,2.088626775,1],
           [6.922596716,1.77106367,1],
           [8.675418651,-0.242068655,1],
           [7.673756466,3.508563011,1]]
neural_net = [[{'weights': [-1.482313569067226,
    1.8308790073202204, 1.078381922048799]},
    {'weights': [0.23244990332399884,
    0.3621998343835864, 0.40289821191094327]}],
     [{'weights': [2.5001872433501404,
        0.7887233511355132, -1.1026649757805829]},
        {'weights': [-2.429350576245497,
        0.8357651039198697, 1.0699217181280656]}]]
for row in dataset:
    prediction = predict(neural_net, row)
    print('Expected=%d, Got=%d' % (row[-1], prediction))
Expected=0, Got=0
Expected=0, Got=0
Expected=0, Got=0
Expected=0, Got=0
Expected=0, Got=0
Expected=1, Got=1
Expected=1, Got=1
Expected=1, Got=1
Expected=1, Got=1
Expected=1, Got=1
```

References

1. Shinde, P. P., & Shah, S. (2018, August). A review of machine learning and deep learning applications. In *2018 Fourth international conference on computing communication control and automation (ICCUBEA)* (pp. 1–6). IEEE.
2. Goh, G., Cammarata, N., Voss, C., Carter, S., Petrov, M., Schubert, L., ... & Olah, C. (2021). Multimodal neurons in artificial neural networks. *Distill*, 6(3), e30.
3. Haoxiang, W., & Smys, S. (2021). Overview of configuring adaptive activation functions for deep neural networks-a comparative study. *Journal of Ubiquitous Computing and Communication Technologies (UCCT)*, 3(01), 10–22.

Exercises

1. What are the steps involved in the training of a neural network?
2. Explain the network initialization step of the neural network along with a diagram.
3. Explain forward propagation along with the related formulae and take the activation function as the sigmoid activation function.
4. Write the pseudocode for forward propagation along with the corresponding functions to apply the activation function. (Take sigmoid as the activation function.)
5. Explain the process of error back propagation through equations by taking the sigmoid activation function.

4

Deep Learning Platforms and Cloud

4.1 OpenCV

OpenCV (Open Source Computer Vision Library [1]) is one of the open source library for machine learning and computer vision. It is used to accelerate the use of machine learning in computer vision products. The code library is easy to use and modify as it is based on the BSD license. It includes state-of-art methods of computer vision and machine learning and more than 2,500 optimized algorithms. The algorithms can be used for face recognition, object recognition, action recognition, camera movement, 3D model recognition, image stitching, eye movement and others. It is supported in Windows, Linux, Android and MacOS, with interfaces of C++, Python, Java and MatLab. Recently, it has also developed interfaces with OpenCL and CUDA. It works well with STL containers and is written in C++. A brief architecture with the modules in OpenCV is as shown in Figure 4.1.

- **Core functionality:** It includes the data structures, multidimensional arrays and basic functions of math that are used by the other modules.
- **Image processing:** It includes various image processing functions like filtering, transformations (resize, affine, remapping), color correction and conversion, histograms and so on.
- **Video analysis:** It includes motion analysis, motion estimation, object tracking, background colors and others.
- **Camera calibration and 3D reconstruction (calib3d):** It consists of different modules for camera movement, multiple views, camera calibration, pose estimation, 3D reconstruction, etc.
- **2D features framework (features2d):** It consists of the descriptors and feature detectors for object recognition applications.
- **Object detection (objdetect):** It consists of modules that can be used for the identification of objects, human–computer interaction along with other frameworks like Caffe and Theano.

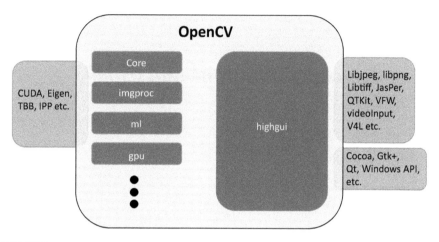

FIGURE 4.1
OpenCV architecture and components.

- **High-level GUI (highgui):** It consists of modules that can be used for building GUI with computer vision applications.
- **Video I/O (videoio):** It consists of modules that are helpful for the codecs in video processing.

An example of the OpenCV program to read an image and convert into gray scale is shown below. Initially, the image is read using the imread() function. The function cvtColor() is used to convert it into gray-scale image.

```
from google.colab.patches import cv2_imshow
# importing libraries
import cv2
import numpy as np
from skimage import io
from PIL import Image
import matplotlib.pylab as plt

image = io.imread("https://thumbor.granitemedia.com/
    yorkshire-terriers-are-a-small-dog-breed/pEK-
    0y844ctqnudyT7emshiviMk=/800x600/filters:format(webp):
    quality(80)/granite-web-prod/ce/39/ce39903185b44964b9d
    b839467e92260.jpg")

cv2_imshow(image)
cv2.waitKey(0)
```

```
gray_image = cv2.cvtColor(image, cv2.COLOR_BGR2GRAY)

cv2_imshow(gray_image)
cv2.waitKey(0)
```

Output

4.2 PyTorch

PyTorch [2] is an open source library for implementing programs on computer vision and multi-modal data. It supports all operating systems with a primary focus on the interface of Python. There are a wide number of applications built on PyTorch, which include Tesla Autopilot, Uber's Pyro, Hugging Face's Transformers, PyTorch Lightning and Catalyst.

The major features of PyTorch are listed as follows.

- **Interface:** It offers easy and usable interface that runs on Python. It uses the services of Python and its framework in a useful way for implementation.
- **Computational graphs:** It uses the computational graph as the background for the execution of tasks. It also gives the option of changing it during run time dynamically.
- **Abstractions:** It provides three levels of abstraction, namely tensor, variable and module. Tensor is used for the creation of n-dimensional array, variable is used for the computational graph, and module is used for storing the learnable weights.

PyTorch offers numerous options for the implementation of various machine learning and deep learning modules. We discuss some of the examples related to PyTorch next.

4.2.1 PyTorch Examples

4.2.1.1 Tensors

The basic entity of PyTorch [3] is the tensor. It provides the option of creating *n*-dimensional arrays as required for the application. The code snippet below demonstrates the creation of 1-D, 2-D and 3-D tensors. The tensor is created using the Tensor() function. In the example, 1-D, 2-D and 3-D tensors are created using oned_data, twod_data and threed_data variables.

```
import torch

#Creating a 1-D tensor
oned_data=[2.0,4.0,6.0]
oned_tensor=torch.Tensor(oned_data)

print(" 1-D Tensor")
print(oned_tensor)

#Creating a 2-D tensor
twod_data=[[1.0,2.0,3.0],[2.0,4.0,6.0]]
twod_tensor=torch.Tensor(twod_data)

print("\n\n 2-D Tensor")
print(twod_tensor)

#Creating a 3-D tensor
threed_data= [
                [[1.0,2.0], [2.0,4.0]],
                [[3.0,6.0], [4.0,8.0]]
             ]
threed_tensor=torch.Tensor(threed_data)

print("\n\n 3-D Tensor")
print(threed_tensor)
```

Output

```
1-D Tensor
tensor([2., 4., 6.])

2-D Tensor
tensor([[1., 2., 3.],
        [2., 4., 6.]])

3-D Tensor
tensor([[[1., 2.],
         [2., 4.]],

        [[3., 6.],
         [4., 8.]]])
```

The size of the tensors is also essential for the operations on the vectors. The following code snippet shows how to determine the size of the tensors. The function size() is used to find the size of the corresponding tensor.

```
#Creating a 1-D tensor
oned_data=[2.0,4.0,6.0]
oned_tensor=torch.Tensor(oned_data)

print(" 1-D Tensor")
print(oned_tensor)

print("Size of tensor")
print(oned_tensor.size())

#Creating a 2-D tensor
twod_data=[[1.0,2.0,3.0],[2.0,4.0,6.0]]
twod_tensor=torch.Tensor(twod_data)

print("\n\n 2-D Tensor")
print(twod_tensor)

print("Size of tensor")
print(twod_tensor.size())

#Creating a 3-D tensor
threed_data= [
                    [[1.0,2.0],  [2.0,4.0]],
                    [[3.0,6.0],  [4.0,8.0]]
              ]
threed_tensor=torch.Tensor(threed_data)

print("\n\n 3-D Tensor")
print(threed_tensor)

print("Size of tensor")
print(threed_tensor.size())
```

Output

```
1-D Tensor
tensor([2., 4., 6.])
Size of tensor
torch.Size([3])

2-D Tensor
tensor([[1., 2., 3.],
        [2., 4., 6.]])
Size of tensor
torch.Size([2, 3])
```

```
3-D Tensor
tensor([[[1., 2.],
         [2., 4.]],

        [[3., 6.],
         [4., 8.]]])
Size of tensor
torch.Size([2, 2, 2])
```

4.2.1.2 Tensor Operations

The different types of operations that can be used on the tensors are addition, dot product and others. The code snippet below demonstrates the different types of tensor operations on tensors A and B. The operator '+' is used for the addition. The method 'add()' can also be used for the addition purpose. The output of the add() function can also be passed as an argument using the 'out' as one of the parameters in the add() function.

```
A=torch.Tensor([1.0,2.0,3.0])
B=torch.Tensor([2.0,4.0,6.0])

#Addition of tensors
C= A+B
print("Tensor sum is :",C)

#Using add function

C=torch.add(A,B)
print("Tensor sum is :",C)

# using method and providing an output tensor as argument
C = torch.empty(3)
torch.add(A, B, out=C)
print("Tensor sum is :",C)

Tensor sum is : tensor([3., 6., 9.])
Tensor sum is : tensor([3., 6., 9.])
Tensor sum is : tensor([3., 6., 9.])
```

PyTorch also supports the in-place addition of two tensors. The function add() is used with the two tensors for the summation. The code snippet for the demonstration of the in-place addition is shown below.

```
#In place Addition

A = torch.Tensor([ 1., 2., 3. ])
B = torch.Tensor([ 4., 5., 6. ])
```

```
B.add_(A)
print("Sum is :",B)

Sum is : tensor([5., 7., 9.])
```

PyTorch supports the flexibility of the conversion of tensors into numpy arrays. This flexibility will facilitate to switch between the machine learning and deep learning modules. The code snippet for the conversion of the tensor into numpy array is shown below. The function numpy() is used for the conversion of the tensor into numpy.

```
#Converting to numpy array

x=torch.ones(5)
print("Original tensor: ",x)

y=x.numpy()
print("Converted Numpy:",y)

x.add_(2)
print("New tensor :",x)
print("New Numpy array:",y)
```

Output

```
Original tensor:  tensor([1., 1., 1., 1., 1.])
Converted Numpy: [1. 1. 1. 1. 1.]
New tensor : tensor([3., 3., 3., 3., 3.])
New Numpy array: [3. 3. 3. 3. 3.]
```

4.3 TensorFlow and Keras

TensorFlow (TF) and Keras are widely used high-level specifications for the implementation of neural network models. Figure 4.2 describes TensorFlow and Keras modules. TF comprises specialized modules for numerical computations in deep learning. It is a preferred tool in the industry and academia to develop deep learning models and architectures. The unique characteristics of the TF are as follows.

- The model of the design can be implemented using Python and used for the defacto execution.
- Computational graph construction is used before the execution using TF.

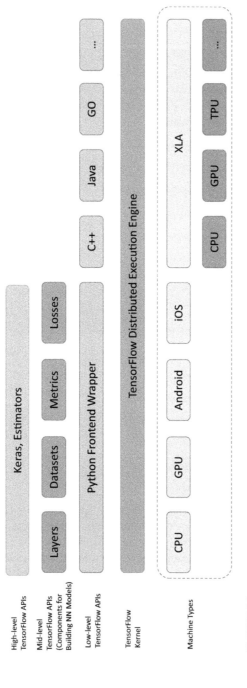

FIGURE 4.2
TensorFlow and Keras modules.

- Ease of training architectures in a distributed way with tensors and flow graphs.

Low-level TensorFlow APIs
The APIs at the low level help in building network graphs that are flexible and tweaked to tune the model. It provides the wrappers around the various classes and operations under the hood of high-level APIs.

Mid-level TensorFlow APIs
It includes a set of packages that are reusable and help in simplifying the process of creation of neural networks. For example, it includes the functions such as layers **(tf.keras.layers)**, Datasets **(tf.data)**, metrics **(tf.keras. metrics)**, loss **(tf.keras.losses)** and FeatureColumns **(tf.feature_column)** packages.

High-level TensorFlow APIs
It provides the API calls in a simplified manner with encapsulation of the details of creating a model. The functions in this level of API have fewer lines of code that help in developing the deep learning models in an easier way.

4.3.1 The Anatomy of a Keras Program

The Keras 'Model' forms the core of a Keras program. A 'Model' is first constructed, then it is compiled. Next, the compiled model is trained and evaluated using their respective training and evaluation datasets. Upon successful evaluation using the relevant metrics, the model is then used for making predictions on previously unseen data samples. Figure 4.3 shows the program flow for modeling with Keras.

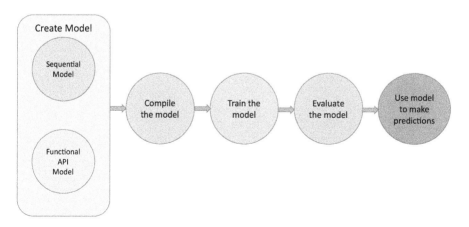

FIGURE 4.3
Keras flow for a program.

We will build a graph to implement Simple Linear Regression. Figure 4.2 represents the graph for the below program. In the below program, we are generating our own data set. There is one dataset without a bias and the other with a bias. During the execution of the program, we can select the one that is needed for us.

```
#importing all the packages

import numpy as np
import tensorflow.compat.v1 as tf
import matplotlib.pyplot as plt

tf.disable_v2_behavior()

#input data without bias:
x_input=np.linspace(0,10,100)
y_input=5*x_input+2.5

#input data with bias:
x_input_bias=np.linspace(0,10,100)
y_input_bias=5*x_input+2.5*np.random.randn(100)

#model parameters
W = tf.Variable(np.random.randn(), name='weight')

#bias
b = tf.Variable(np.random.randn(), name='bias')

#placeholders
with tf.name_scope('input') as scope:
    X=tf.placeholder(tf.float32)
    Y=tf.placeholder(tf.float32)

#model
with tf.name_scope('model') as scope:
    Y_pred=tf.add(tf.multiply(X,W),b)

#loss
with tf.name_scope('loss') as scope:
    loss = tf.reduce_mean(tf.square(Y_pred - Y ))

#record the summary of the cost function
cost=tf.summary.scalar("loss", loss)
```

```
#training algorithm
with tf.name_scope('training') as scope:
    optimizer = tf.train.GradientDescentOptimizer(0.01)
    train = optimizer.minimize(loss)

#initializing the variables
#init = tf.initialize_all_variables() #for TF version < 1.0
init=tf.global_variables_initializer()

#merge all summary into a single operator
merged_summary = tf.summary.merge_all()

#setting the number of epoch
epoch=500
summary_writer=tf.summary.FileWriter('./finallogs',
graph=tf.get_default_graph())

#starting the session

with tf.Session() as sess:
    sess.run(init)

    for step in range(epoch):
        for x,y in zip(x_input,y_input):
            sess.run(train,feed_dict={X:x,Y:y})

        if step%50==0:
            W1 = sess.run(W)
            B1 = sess.run(b)
            c = sess.run(loss,feed_dict={X: x_input, Y:
              y_input})
            print('Epochs %f Cost %f Weight %f Bias %f'
              % (step,c,W1,B1))
    print("Model paramters:")
    print("Weight:%f" %sess.run(W))
    print("bias:%f" %sess.run(b))
```

Output

Here we have two outputs.

1. When there is no bias in the generated data set as shown in Figure 4.4.
2. When there is some bias in the generated data set as shown in Figure 4.5.

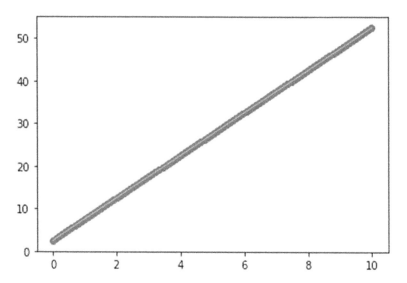

FIGURE 4.4
Basic plot of the data with no bias.

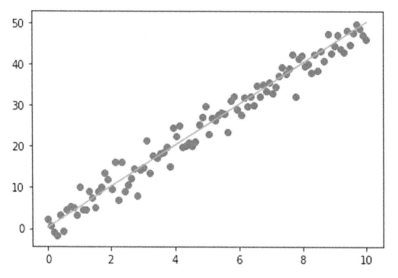

FIGURE 4.5
Basic plot of the data with bias.

The program output with no biased input.

```
Epochs 0.000000 Cost 0.111451 Weight 4.942766 Bias 3.075311
Epochs 50.000000 Cost 0.000000 Weight 5.000000 Bias 2.500004
Epochs 100.000000 Cost 0.000000 Weight 5.000000 Bias 2.500002
Epochs 150.000000 Cost 0.000000 Weight 5.000000 Bias 2.500002
Epochs 200.000000 Cost 0.000000 Weight 5.000000 Bias 2.500002
Epochs 250.000000 Cost 0.000000 Weight 5.000000 Bias 2.500002
Epochs 300.000000 Cost 0.000000 Weight 5.000000 Bias 2.500002
Epochs 350.000000 Cost 0.000000 Weight 5.000000 Bias 2.500002
Epochs 400.000000 Cost 0.000000 Weight 5.000000 Bias 2.500002
Epochs 450.000000 Cost 0.000000 Weight 5.000000 Bias 2.500002
Model paramters:
Weight:5.000000
bias:2.500002
```

The basic plot of the data with no bias.
The program output with biased input.

```
Epochs 0.000000 Cost 8.137952 Weight 4.770332 Bias 2.425493
Epochs 50.000000 Cost 6.503957 Weight 4.985663 Bias 0.261007
Epochs 100.000000 Cost 6.503957 Weight 4.985664 Bias 0.260995
Epochs 150.000000 Cost 6.503957 Weight 4.985664 Bias 0.260995
Epochs 200.000000 Cost 6.503957 Weight 4.985664 Bias 0.260995
Epochs 250.000000 Cost 6.503957 Weight 4.985664 Bias 0.260995
Epochs 300.000000 Cost 6.503957 Weight 4.985664 Bias 0.260995
Epochs 350.000000 Cost 6.503957 Weight 4.985664 Bias 0.260995
Epochs 400.000000 Cost 6.503957 Weight 4.985664 Bias 0.260995
Epochs 450.000000 Cost 6.503957 Weight 4.985664 Bias 0.260995
Model paramters:
Weight:4.985664
bias:0.260995
```

The basic plot of the data with bias.

References

1. Bradski, G., & Kaehler, A. (2000). OpenCV. *Dr. Dobb's Journal of Software Tools*, 3, 2.
2. Paszke, A., Gross, S., Massa, F., Lerer, A., Bradbury, J., Chanan, G., ... & Chintala, S. (2019). Pytorch: An imperative style, high-performance deep learning library. *Advances in Neural Information Processing Systems*, 32.
3. Dillon, J. V., Langmore, I., Tran, D., Brevdo, E., Vasudevan, S., Moore, D., ... & Saurous, R. A. (2017). Tensorflow distributions. arXiv preprint arXiv:1711.10604.

Exercises

1. Describe OpenCV and its applications.
2. Describe the core modules of OpenCV.
3. Describe PyTorch. Explain the major features of PyTorch.
4. Describe some of the tensor operations along with examples using sample function calls and the corresponding output by taking some sample input.
5. What are the unique characteristics of TensorFlow? Explain the various levels of the TensorFlow API along with a neat diagram.

Part II

Multi-Modal Data and Analytics with Neural Networks using Cloud

5

Neural Networks for Multi-Modal Data Analytics

5.1 Introduction

5.1.1 Convolutional Neural Network (CNN)

Convolutional neural networks (CNNs) are a family of neural networks designed for the purpose of image processing, object detection, anomaly detection and other applications [1]. CNN-based architectures are ubiquitous in every field of computer vision domain, such as automobiles, self-driving, molecular biology and other applications. CNNs are computationally efficient as they can be parallelized, and thus computation efficiency can be achieved easily. It requires fewer parameters and fully connected layers to achieve the required level of accuracy as well. CNNs can be used on different types of data like image, audio, text and time series. In this chapter, we introduce the basics of CNN and its operations with relevant examples.

5.1.2 Convolution and Cross-Correlation

In a CNN, there are three main components, namely input, kernel and output [2, 3]. The input and kernel are combined to produce the output (feature map). The input data is generally an image that consists of multiple color channels. In this section, the color channels are ignored and assumed as only a single channel. In this section, let us understand the difference between convolution and cross-correlation. The definition of convolution can be defined on two functions, 's' and 'w', as shown in Equation (5.1). For the discrete objects, the integral turns into a sum, as shown in Equation (5.2).

$$(s*w)(x) = \int s(t)g(x-t)dt \tag{5.1}$$

$$(s*w)(z) = \sum_i s(i)g(z-i) \tag{5.2}$$

DOI: 10.1201/9781003215974-7

Convolutions are generally used on two dimensions of the image, as shown in Equation (5.3). Cross-correlation is different from convolution where the kernels are not flipped, as shown in Equation (5.4). In this entire chapter of CNN, we use Equation (5.4) as the basis for the operations in CNN.

$$(s * w)(z, x) = \sum_i \sum_j s(i, j) g(z - i)(x - j) \qquad (5.3)$$

$$(s * w)(z, x) = \sum_i \sum_j s(i, j) g(z + i)(x + j) \qquad (5.4)$$

Let us consider an example for the demonstration of the convolution as shown in Figure 5.1. Here, the input size considered is 3 × 3, kernel is 2 × 2 and the output is 2 × 2. The convolution operations begin from the top-left corner of the input and continue to the right bottom. The convolution window slides from one stride to another to find the scalar output value. The output size is determined using Equation (5.5), where I_h and I_w are the input height and width, K_h and K_w are the kernel height and kernel width, respectively, and $F_{h \times w}$ denotes the height and width of the output. Therefore, for the

Input			Kernel		Output	
5	3	2	0	2	21	14
3	0	10	5	1	62	113
9	17	8				
5	3	2	0	2	21	14
3	0	10	5	1	62	113
9	17	8				
5	3	2	0	2	21	14
3	0	10	5	1	62	113
9	17	8				
5	3	2	0	2	21	14
3	0	10	5	1	62	113
9	17	8				

FIGURE 5.1
Demonstration of elements of the output of the convolution.

considered example, the output size is 2 × 2, as shown in Equation (5.6). The four elements of the output are derived as follows for the example shown in Figure 5.1.

$$F_{h\times w} = (I_h - K_h + 1) \times (I_w - K_w + 1) \tag{5.5}$$

$$F_{h\times w} = (3 - 2 + 1) \times (3 - 2 + 1) = 2 \times 2 \tag{5.6}$$

The code implementation for the considered example is shown below. The function corr2d is used to get the feature map output that takes the input and the kernel as the arguments. The shape method determines the size of the input and the kernel. The output size is derived using Equation (5.5). The final values of the output are derived using Equation (5.4). In this way, the basic convolution is operated on a given input and kernel.

```
import tensorflow as tf
def corr2d(X, K):
    """Compute 2D cross-correlation."""
    h, w = K.shape
    Y = tf.Variable(tf.zeros((X.shape[0] - h + 1,
        X.shape[1] - w + 1)))
    for i in range(Y.shape[0]):
        for j in range(Y.shape[1]):
            Y[i, j].assign(tf.reduce_sum(X[i:i + h, j:j +
                w] * K))
    return Y
X = tf.constant([[5.0, 3.0, 2.0], [3.0, 0.0, 10.0], [9.0,
17.0, 8.0]])
K = tf.constant([[0.0, 2.0], [5.0, 1.0]])
corr2d(X, K)
```

Output

```
<tf.Variable 'Variable:0' shape=(2, 2) dtype=float32,
    numpy=
array([[ 21., 14.],
       [ 62., 113.]], dtype=float32)>
```

5.1.2.1 Padding and Stride

In the previous section, the example considered had the same dimensions of height and width for input and kernel. However, several cases of CNN dealing with image processing applications require padding and stride convolutions. For example, if we consider an image of size 240 × 240, applying ten layers of 5 × 5 convolutions will reduce the image size by 200 × 200. This reduction in size will not capture the necessary information present in the boundaries of the images. Padding and strides help in producing the feature map outputs that are feature specific to some applications. In this section, we explain the usage of padding and stride in convolution neural networks.

Padding

The usage of small kernels for convolutions results in the loss of information of the image boundaries and other pixels. The solution to avoid such problems is to add extra pixels to fill the boundary of image and produce the feature map output that is the same as the size of the input. The size of the output with padding is determined as shown in Equations (5.7) and (5.8), where the extra terms P_h and P_w refer to the padding height and padding width.

Let us consider an example as shown in the figure. Here, the input of size 3×3 is padded to increase its size to 5×5. The shaded portions show the padding added to the input. The derivation of the values of the feature map output is the same as the convolutions defined in the previous sections. For, the considered example, the output size is 4×4, as shown in Equation (5.8). The four elements of the output are derived as follows for the example shown in Figure 5.2.

$$F_{h \times w} = \left(I_h - K_h + P_h + 1 \right) \times \left(I_w - K_w + P_w + 1 \right) \tag{5.7}$$

$$F_{h \times w} = \left(3 - 2 + 2 + 1 \right) \times \left(3 - 2 + 2 + 1 \right) \tag{5.8}$$

The kernels used in the CNNs are usually of odd sizes, such as 1, 3, 5 or 7. It preserves the spatial dimensions of the image when padding is used to derive the feature map output. For any 2-D image, the odd-sized kernel and padding on all the sides of the image produce the output of the same size as the input and thereby preserves the spatial dimensions of the image. The code snippet demonstrating the usage of padding is shown below.

```
import tensorflow as tf
def corr2d(X, K):
    """Compute 2D cross-correlation."""
    h, w = K.shape
    Y = tf.Variable(tf.zeros((X.shape[0] - h + 1,
        X.shape[1] - w + 1)))
    for i in range(Y.shape[0]):
        for j in range(Y.shape[1]):
            Y[i, j].assign(tf.reduce_sum(X[i:i + h, j:j +
                w] * K))
    return Y
X = tf.constant([[0.0, 0.0, 0.0,0.0,0.0],[0.0,5.0, 3.0,
    2.0,0.0], [0.0,3.0, 0.0, 10.0,0.0], [0.0,9.0, 17.0,
    8.0,0.0],[0.0, 0.0, 0.0,0.0,0.0]])
K = tf.constant([[0.0, 2.0], [5.0, 1.0]])
corr2d(X, K)
```

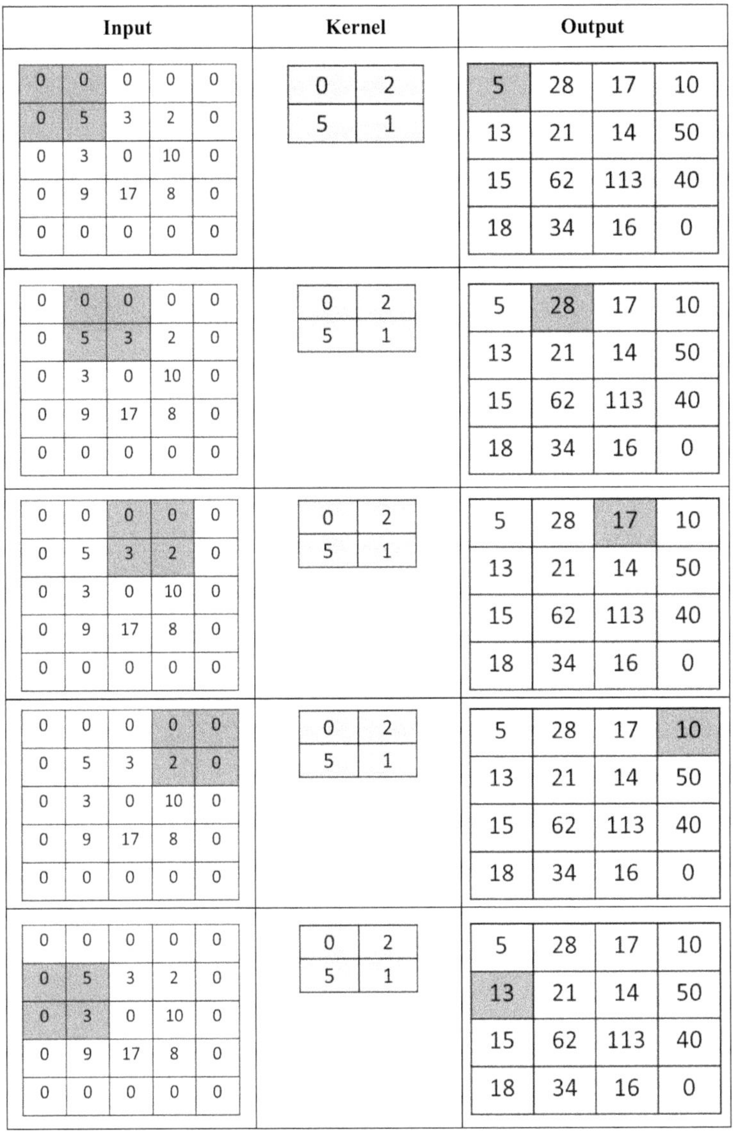

Input					Kernel		Output			
0	0	0	0	0	0	2	5	28	17	10
0	5	3	2	0	5	1	13	21	14	50
0	3	0	10	0			15	62	113	40
0	9	17	8	0			18	34	16	0
0	0	0	0	0						

Input					Kernel		Output			
0	0	0	0	0	0	2	5	28	17	10
0	5	3	2	0	5	1	13	21	14	50
0	3	0	10	0			15	62	113	40
0	9	17	8	0			18	34	16	0
0	0	0	0	0						

Input					Kernel		Output			
0	0	0	0	0	0	2	5	28	17	10
0	5	3	2	0	5	1	13	21	14	50
0	3	0	10	0			15	62	113	40
0	9	17	8	0			18	34	16	0
0	0	0	0	0						

Input					Kernel		Output			
0	0	0	0	0	0	2	5	28	17	10
0	5	3	2	0	5	1	13	21	14	50
0	3	0	10	0			15	62	113	40
0	9	17	8	0			18	34	16	0
0	0	0	0	0						

Input					Kernel		Output			
0	0	0	0	0	0	2	5	28	17	10
0	5	3	2	0	5	1	13	21	14	50
0	3	0	10	0			15	62	113	40
0	9	17	8	0			18	34	16	0
0	0	0	0	0						

FIGURE 5.2
Demonstrating the process of padding. *(Continued)*

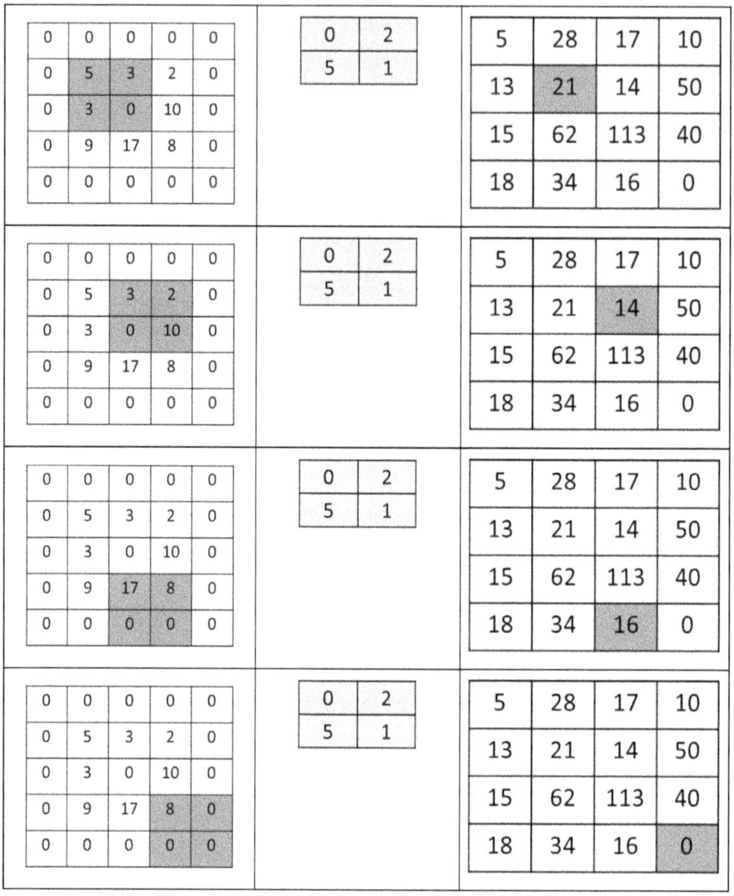

0	0	0	0	0
0	5	3	2	0
0	3	0	10	0
0	9	17	8	0
0	0	0	0	0

0	2
5	1

5	28	17	10
13	21	14	50
15	62	113	40
18	34	16	0

0	0	0	0	0
0	5	3	2	0
0	3	0	10	0
0	9	17	8	0
0	0	0	0	0

0	2
5	1

5	28	17	10
13	21	14	50
15	62	113	40
18	34	16	0

0	0	0	0	0
0	5	3	2	0
0	3	0	10	0
0	9	17	8	0
0	0	0	0	0

0	2
5	1

5	28	17	10
13	21	14	50
15	62	113	40
18	34	16	0

0	0	0	0	0
0	5	3	2	0
0	3	0	10	0
0	9	17	8	0
0	0	0	0	0

0	2
5	1

5	28	17	10
13	21	14	50
15	62	113	40
18	34	16	0

FIGURE 5.2   (CONTINUED)

Output

```
<tf.Variable 'Variable:0' shape=(4, 4) dtype=float32,
numpy=
array([[  5.,   28.,   17.,   10.],
       [ 13.,   21.,   14.,   50.],
       [ 15.,   62.,  113.,   40.],
       [ 18.,   34.,   16.,    0.]], dtype=float32)>
```

Stride

In the previous sections, the feature map output was obtained starting at the top-left corner of the input and then sliding it one element at a type. For more computational efficiency, the window is moved with more than one element at a time. Figure 5.3 demonstrates the operation of stride. Stride refers to rows and columns traversed. The output size is determined using the stride for the convolution as shown in Equation (5.9).

$$F_{h \times w} = \left\lfloor (I_h - K_h + P_h + S_h + 1) \times (I_w - K_w + P_w + S_w + 1) \right\rfloor \qquad (5.9)$$

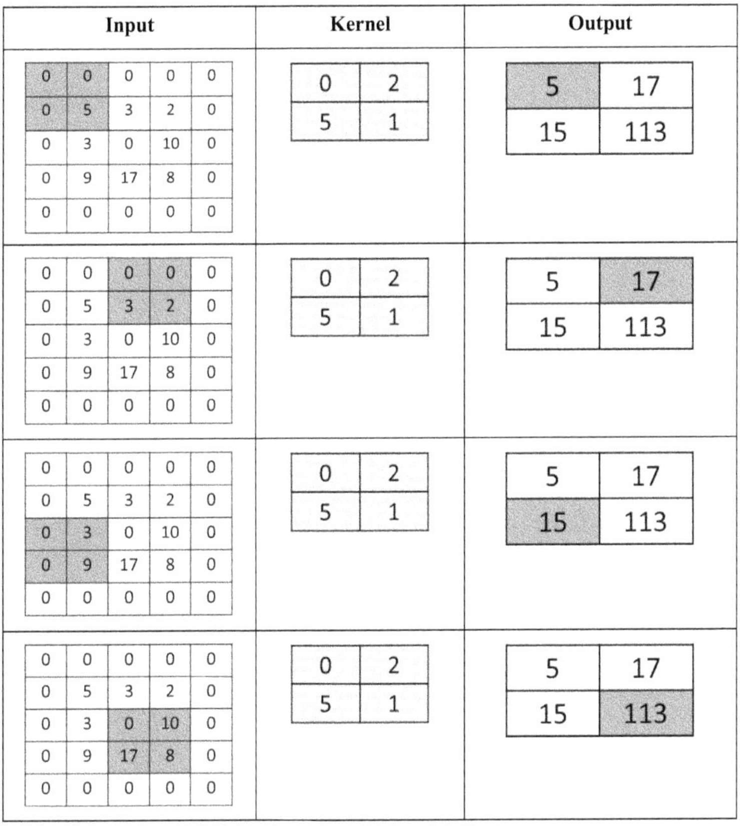

Input					Kernel		Output	
0	0	0	0	0	0	2	5	17
0	5	3	2	0	5	1	15	113
0	3	0	10	0				
0	9	17	8	0				
0	0	0	0	0				

Input					Kernel		Output	
0	0	0	0	0	0	2	5	17
0	5	3	2	0	5	1	15	113
0	3	0	10	0				
0	9	17	8	0				
0	0	0	0	0				

Input					Kernel		Output	
0	0	0	0	0	0	2	5	17
0	5	3	2	0	5	1	15	113
0	3	0	10	0				
0	9	17	8	0				
0	0	0	0	0				

Input					Kernel		Output	
0	0	0	0	0	0	2	5	17
0	5	3	2	0	5	1	15	113
0	3	0	10	0				
0	9	17	8	0				
0	0	0	0	0				

FIGURE 5.3
Demonstrating the operation of the stride.

5.1.3 A Simple CNN Example

A simple CNN with TensorFlow and Keras is explained in this section. The dataset used for constructing the CNN is MNIST. This is a dataset of 60,000 28 × 28 grayscale images of ten digits, along with a test set of 10,000 images. The overall structure of the CNN that will be implemented is shown in Figure 5.4. There are four convolutional layers of 50 filters, each 3 × 3 in size, two max-pooling layers and two dense layers. One of the dense layers is used for flattening, and the other is used with the softmax layer for classification.

The dataset is first divided into two sets, namely training and testing. The function reshape() is first used for reshaping the training and testing data so that all the images fall into the size 28 × 28. The categories of ten digits are made using the one-hot encoding scheme using the function to_categorical. The shape of the training and testing is validated first to check if the number of images with the description matches.

The function Sequential() is used to build the model of the CNN. The convolutional layers are added using this function. The first layer with 50 filters of 3 × 3 size is added, followed by the second layer and max-pooling layer. In this way, a sequential CNN is implemented with four convolutional layers

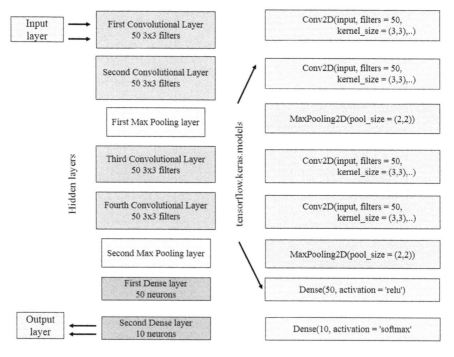

FIGURE 5.4
Structure of the CNN.

and two max-pooling layers. At the end, two dense layers one with relu and softmax functions are implemented for the final classification.

The function summary() is used to determine the total number of parameters in the CNN and each layer. The fit() function is used for the training of the entire dataset and for printing the accuracy for each epoch. It can be seen that the overall accuracy of the entire CNN is 98.97%. The prediction function is used for prediction purposes in the end. In this way, a CNN can be modeled sequentially with the required number of layers.

```python
import numpy as np
import matplotlib.pyplot as plt

from tensorflow.keras.datasets import mnist
from tensorflow.keras.utils import to_categorical

from tensorflow.keras.models import Sequential
from tensorflow.keras import optimizers
from tensorflow.keras.layers import Dense, Activation,
Flatten, Conv2D, MaxPooling2D

(X_train, y_train), (X_test, y_test) = mnist.load_data()
# reshaping X data: (n, 28, 28) => (n, 28, 28, 1)
X_train = X_train.reshape((X_train.shape[0], X_train.
    shape[1], X_train.shape[2], 1))
X_test = X_test.reshape((X_test.shape[0], X_test.
    shape[1], X_test.shape[2], 1))

# converting y data into categorical (one-hot encoding)
y_train = to_categorical(y_train)
y_test = to_categorical(y_test)

print(X_train.shape)
print(X_test.shape)
print(y_train.shape)
print(y_test.shape)

(60000, 28, 28, 1)
(10000, 28, 28, 1)
(60000, 10)
(10000, 10)

def deep_cnn():
    deep_cnn_model = Sequential()

    deep_cnn_model.add(Conv2D(input_shape = (X_train.
        shape[1], X_train.shape[2], X_train.shape[3]),
        filters = 50, kernel_size = (3,3), strides =
        (1,1), padding = 'same'))
```

```
deep_cnn_model.add(Activation('relu'))
deep_cnn_model.add(Conv2D(filters = 50, kernel_size =
    (3,3), strides = (1,1), padding = 'same'))
deep_cnn_model.add(Activation('relu'))
deep_cnn_model.add(MaxPooling2D(pool_size = (2,2)))
deep_cnn_model.add(Conv2D(filters = 50, kernel_size =
    (3,3), strides = (1,1), padding = 'same'))
deep_cnn_model.add(Activation('relu'))
deep_cnn_model.add(Conv2D(filters = 50, kernel_size =
    (3,3), strides = (1,1), padding = 'same'))
deep_cnn_model.add(Activation('relu'))
deep_cnn_model.add(MaxPooling2D(pool_size = (2,2)))

# prior layer should be flattend to be connected to dense
layers
deep_cnn_model.add(Flatten())
# dense layer with 50 neurons
deep_cnn_model.add(Dense(50, activation = 'relu'))
# final layer with 10 neurons to classify the instances
deep_cnn_model.add(Dense(10, activation = 'softmax'))

adam = optimizers.Adam(lr = 0.001)
deep_cnn_model.compile(loss = 'categorical_crossentropy',
    optimizer = adam, metrics = ['accuracy'])

    return deep_cnn_model

model = deep_cnn()
model.summary()

Model: "sequential_13"
```

Layer (type)	Output Shape	Param #
conv2d_85 (Conv2D)	(None, 28, 28, 50)	500
activation_78 (Activation)	(None, 28, 28, 50)	0
conv2d_86 (Conv2D)	(None, 28, 28, 50)	22550
activation_79 (Activation)	(None, 28, 28, 50)	0
max_pooling2d_39 (MaxPooling	(None, 14, 14, 50)	0
conv2d_87 (Conv2D)	(None, 14, 14, 50)	22550
activation_80 (Activation)	(None, 14, 14, 50)	0

```
conv2d_88 (Conv2D)              (None, 14, 14, 50)   22550

activation_81 (Activation)      (None, 14, 14, 50)   0

max_pooling2d_40 (MaxPooling)   (None, 7, 7, 50)     0

flatten_9 (Flatten)             (None, 2450)         0

dense_18 (Dense)                (None, 50)           122550

dense_19 (Dense)                (None, 10)           510
=================================================================
Total params: 191,210
Trainable params: 191,210
Non-trainable params: 0


%%time
history = model.fit(X_train, y_train, batch_size = 100,
validation_split = 0.2, epochs = 5, verbose = 1)

Epoch 1/5
480/480 [==============================] - 301s 627ms/
    step - loss: 0.0230 - accuracy: 0.9930 - val_loss:
    0.0478 - val_accuracy: 0.9869
Epoch 2/5
480/480 [==============================] - 301s 626ms/
    step - loss: 0.0155 - accuracy: 0.9955 - val_loss:
    0.0429 - val_accuracy: 0.9880
Epoch 3/5
480/480 [==============================] - 299s 623ms/
    step - loss: 0.0153 - accuracy: 0.9950 - val_loss:
    0.0444 - val_accuracy: 0.9877
Epoch 4/5
480/480 [==============================] - 298s 622ms/
    step - loss: 0.0144 - accuracy: 0.9953 - val_loss:
    0.0521 - val_accuracy: 0.9880
Epoch 5/5
480/480 [==============================] - 299s 623ms/
    step - loss: 0.0183 - accuracy: 0.9940 - val_loss:
    0.0441 - val_accuracy: 0.9897
CPU times: user 47min 7s, sys: 43.8 s, total: 47min 51s
Wall time: 25min 22s
```

5.1.4 Recurrent Neural Network (RNN)

A recurrent neural network (RNN) is a type of neural network that is suitable for time series analysis or the data that involves sequences [4]. The feed-forward neural networks are suitable for the data points that are independent of each other. If there is a sequence of the points that need to be used for the classification and prediction, then RNN can be used. They use memory gates to store the previous information and generate the next outputs in the sequence. In this section, we explain the structure of the RNN and its variants with a simple implementation of the RNN.

5.1.4.1 Types of Recurrent Neural Network

A simple RNN structure is shown in Figure 5.5. It consists of the input layer 'x', hidden layer 'H' and the output layer 'y'. The recurrent connections exist between the hidden units. The right side of the figure shows the unfolding of RNN over the time steps 't'. The recurrent connections exist between the hidden node at $t-1 \rightarrow t$ and $t- > t+1$. These recurrent connections use the common weight matrix 'W' for the computations.

The various types of RNN are shown in Figure 5.6. Figure (a) shows the vanilla RNN, that is, one-to-one. Figure (b) shows the type of one-to-many-variant RNN with one input and multiple outputs. Figure (c) shows the many-to-one-variant RNN with many inputs and one output. Figure (d)

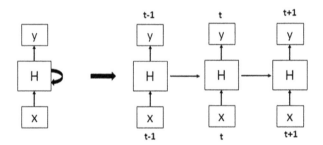

FIGURE 5.5
RNN structure.

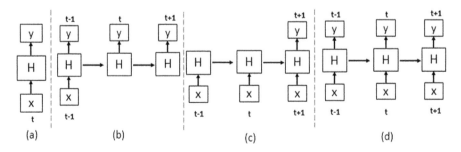

FIGURE 5.6
Types of RNN. (a) One-to-one (b) One-to-many (c) Many-to-one (d) Many-to-many.

shows the many-to-many-variant RNN with many inputs and many out-puts. All the variants here are displayed with the time series as 't'.

5.1.4.2 RNN with Keras on Dataset

In this section, we discuss the implementation of a simple RNN. The dataset that is taken for the implementation is Reuters. This is a dataset of 11,228 newswires from Reuters, labeled over 46 topics. Initially, the dataset is loaded with the parameters as num_words, max_length and the split ratio. The labels of the data are extracted using the function load_data() with the given parameters. It is followed by the label association with the training and testing set.

The function Sequential() is used in the Keras for the implementation of the RNN layers. The SimpleRNN() function is implemented with 50 neu-ron units and two dense layers of 46 units. The accuracy and loss values are printed in each epoch with an overall accuracy of 72%. The loss function cat-egorical cross entropy is used for estimation and visualized at the end using the matplotlib. The outputs of model accuracy and model loss are shown in Figures 5.7 and 5.8.

```
import numpy as np
from sklearn.metrics import accuracy_score
from tensorflow.keras.datasets import reuters
from tensorflow.keras.preprocessing.sequence import
    pad_sequences
from tensorflow.keras.utils import to_categorical
num_words = 30000
max_length = 50
train_test_split = 0.3
```

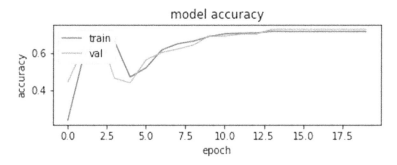

FIGURE 5.7
Model accuracy.

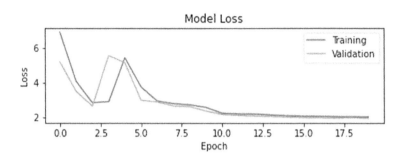

FIGURE 5.8
Model loss.

```
(train_x, train_labels), (test_x, test_labels) = reuters.
   load_data(num_words = num_words, maxlen = max_length,
   test_split = train_test_split)

train_x = pad_sequences(train_x, padding = 'post')
test_x = pad_sequences(test_x, padding = 'post')

train_x = np.array(train_x).reshape((train_x.shape[0],
train_x.shape[1], 1))

test_x = np.array(test_x).reshape((test_x.shape[0],
test_x.shape[1], 1))

labels = np.concatenate((train_labels, test_labels))
labels = to_categorical(labels)

train_labels = labels[:1395]
test_labels = labels[1395:]

print(train_x.shape)
print(test_x.shape)
print(train_labels.shape)
print(test_labels.shape)

(1395, 49, 1)
(599, 49, 1)
(1395, 46)
(599, 46)

from tensorflow.keras.models import Sequential
from tensorflow.keras.layers import Dense, Activation,
   SimpleRNN
from tensorflow.keras import optimizers
```

```
from tensorflow.keras.wrappers.scikit_learn import
    KerasClassifier

def RNN():
    model = Sequential()
    model.add(SimpleRNN(50, input_shape = (49,1), return_
        sequences = False))
    model.add(Dense(46))
    model.add(Dense(46))
    model.add(Activation('relu'))
    adam = optimizers.Adam(lr = 0.001)
    model.compile(loss = 'categorical_crossentropy',
        optimizer = adam, metrics = ['acc'])

    return model
model = KerasClassifier(build_fn = gru, epochs = 20,
    batch_size = 50, verbose = 1)

history = model.fit(train_x, train_labels,
    validation_split=0.2)

Epoch 1/20
23/23 [==============================] - 1s 26ms/step -
    loss: 6.9169 - acc: 0.2446 - val_loss: 5.1993 - val_
    acc: 0.4480
Epoch 2/20
23/23 [==============================] - 0s 16ms/step -
    loss: 4.0892 - acc: 0.5815 - val_loss: 3.5060 - val_
    acc: 0.6237
Epoch 3/20
23/23 [==============================] - 0s 13ms/step -
    loss: 2.8365 - acc: 0.6891 - val_loss: 2.6424 - val_
    acc: 0.6989
Epoch 4/20
23/23 [==============================] - 0s 14ms/step -
    loss: 2.8963 - acc: 0.6613 - val_loss: 5.5496 - val_
    acc: 0.4659
Epoch 5/20
23/23 [==============================] - 0s 13ms/step -
    loss: 5.4360 - acc: 0.4722 - val_loss: 5.1489 - val_
    acc: 0.4409
Epoch 6/20
23/23 [==============================] - 0s 14ms/step -
    loss: 3.7467 - acc: 0.5197 - val_loss: 2.9740 - val_
    acc: 0.5627
Epoch 7/20
23/23 [==============================] - 0s 13ms/step -
    loss: 2.9358 - acc: 0.6156 - val_loss: 2.8750 - val_
    acc: 0.6022
```

```
Epoch 8/20
23/23 [==============================] - 0s 14ms/step -
   loss: 2.7803 - acc: 0.6478 - val_loss: 2.6417 - val_
   acc: 0.6201
Epoch 9/20
23/23 [==============================] - 0s 13ms/step -
   loss: 2.7146 - acc: 0.6622 - val_loss: 2.6063 - val_
   acc: 0.6416
Epoch 10/20
23/23 [==============================] - 0s 15ms/step -
   loss: 2.5680 - acc: 0.6864 - val_loss: 2.3583 - val_
   acc: 0.6882
Epoch 11/20
23/23 [==============================] - 0s 13ms/step -
   loss: 2.2329 - acc: 0.7007 - val_loss: 2.1579 - val_
   acc: 0.6882
Epoch 12/20
23/23 [==============================] - 0s 14ms/step -
   loss: 2.1981 - acc: 0.7043 - val_loss: 2.1119 - val_
   acc: 0.6989
Epoch 13/20
23/23 [==============================] - 0s 13ms/step -
   loss: 2.1846 - acc: 0.7052 - val_loss: 2.0693 - val_
   acc: 0.6989
Epoch 14/20
23/23 [==============================] - 0s 14ms/step -
   loss: 2.1323 - acc: 0.7133 - val_loss: 2.0438 - val_
   acc: 0.7240
Epoch 15/20
23/23 [==============================] - 0s 13ms/step -
   loss: 2.0927 - acc: 0.7124 - val_loss: 2.0061 - val_
   acc: 0.7240
Epoch 16/20
23/23 [==============================] - 0s 13ms/step -
   loss: 2.0642 - acc: 0.7124 - val_loss: 1.9806 - val_
   acc: 0.7240
Epoch 17/20
23/23 [==============================] - 0s 14ms/step -
   loss: 2.0549 - acc: 0.7124 - val_loss: 1.9669 - val_
   acc: 0.7240
Epoch 18/20
23/23 [==============================] - 0s 13ms/step -
   loss: 2.0377 - acc: 0.7124 - val_loss: 1.9464 - val_
   acc: 0.7240
Epoch 19/20
23/23 [==============================] - 0s 13ms/step -
   loss: 2.0261 - acc: 0.7124 - val_loss: 1.9811 - val_
   acc: 0.7240
```

```
Epoch 20/20
23/23 [==============================] - 0s 13ms/step -
    loss: 2.0196 - acc: 0.7124 - val_loss: 1.9289 - val_
    acc: 0.7240

pred_labels = model.predict(test_x)
test_labels_ = np.argmax(test_labels, axis = 1)
print("Test accuracy: "+str(accuracy_score(pred_labels,
    test_labels_)))

Test accuracy: 0.7479131886477463
from matplotlib import pyplot as plt

plt.subplot(211)
plt.plot(history.history['acc'])
plt.plot(history.history['val_acc'])
plt.title('model accuracy')
plt.ylabel('accuracy')
plt.xlabel('epoch')
plt.legend(['train', 'val'], loc='upper left')
plt.show()

plt.subplot(212)
plt.plot(history.history['loss'])
plt.plot(history.history['val_loss'])
plt.title('Model Loss')
plt.ylabel('Loss')
plt.xlabel('Epoch')
plt.legend(['Training', 'Validation'], loc='upper right')

plt.tight_layout()
plt.show()
```

5.1.5 Long Short-Term Memory (LSTM)

Sequence-to-sequence problems can be solved using RNNs for short-term dependencies [4]. It cannot be used for problems to arrive at decisions based on the context. For example, consider the prediction for the sentence 'The color of the leaf is ___'. Here, RNN needs to remember the previous context and in most cases leaf is green in color. Therefore, the output would be 'The color of the leaf is green'. However, consider the prediction for the sentence 'I spent 10 years in Russia. I then moved to United States. I can speak fluent ____'. The obvious answer to fill in the gap is Russian. Here, the RNN needs to remember the context for the prediction. There is large information that RNN has to store for the prediction.

In a basic feed-forward neural network, the weight update is the combination of learning rate, error term and the previous layer input. The error term is dependent on the product of all the previous layers. The derivatives of the activation functions get multiplied various times as it moves to the starting layers. Hence, the gradient vanishes and becomes difficult for training. LSTM eliminates such long-term dependencies with the use of memory gate for sequence-to-sequence prediction problems. In this section, we discuss the LSTM architecture, components and related examples.

5.1.5.1 LSTM Architecture

A typical LSTM network consists of memory cells as shown in the figure. Memory cells are responsible for managing the information in the network with cell state and hidden state. The three important components of an LSTM network are input gate, output gate and memory gate. Each of these gates is discussed as follows. An example of the LSTM network is shown in Figure 5.9. The cell in the LSTM memory block has the input gate i_t, memory block m_t and a forget gate g_t. The forget gate g_t is another gate present whose function is to select the parts of the long-term memory, that is, c_{t-1}, to be remembered.

The computation of LSTM from an input sequence x to the output sequence y is calculated using the following Equations (5.10)–(5.12), where W terms are weight matrices (e.g., W_{ix} is the weights from the input gate to input), b denotes the bias vectors, σ denotes the sigmoid as the activation function and .(dot) denotes the dot product of the vectors. The terms i, g, o and c represent the input gate, forget gate, output gate and the cell state of the LSTM block. The input and the output functions are represented using f and g functions.

$$f = \sigma \left(W x_t + W m_{t-1} + W C_{t-1} \right) + b \tag{5.10}$$

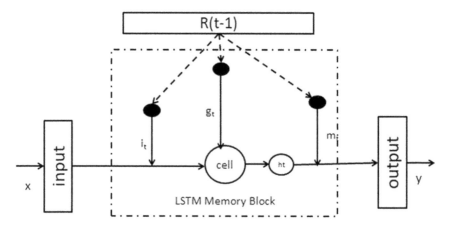

FIGURE 5.9
LSTM architecture.

$$c = f.c_{t-1} + i_t.f\left(Wx_t + b\right) \tag{5.11}$$

$$y = \Pi\left(Wg\left(m_t\right) + b\right) \tag{5.12}$$

Forget Gate
Forget gate removes the information from the cell state. The LSTM network uses the memory gate to remove the information from the network with the usage of a filter. It takes the input from the previous hidden state ht_{-1} and input at the time t x_t. Suppose the activation function used is sigmoid in the LSTM network, the cell state gives the output ranging from the values of 0 and 1. A value of '0' indicates that the information needs to be discarded, whereas the value of '1' indicates that the information needs to be stored. The final vector from the sigmoid activation function is multiplied to a cell state.

For example, consider the following sentence, 'Rao is a nice person. Ananth is evil in nature'. The LSTM network forgets the context as soon as it encounters the first full stop after the person. The forget gate uses the second subject 'Ananth' to replace the necessary information as required in the context.

Input Gate
The input gate in the LSTM network is used for handling the additional information of the cell state. It uses the value of the sigmoid function for the forget gate to provide additional information for regulation. It creates a vector with the activation function 'tanh' using the information from the previous hidden cell state values. The vector is then multiplied with the previous hidden state value as an additional information.

For example, consider the sentence 'Rao knows shooting. He told me over the conversation he has served navy in the past'. The important information from the forget gate received will be 'Rao knows shooting'. The input gate saves the cell state information of serving the navy along with shooting.

Output Gate
It creates a vector using the tanh as the activation function. A range of values from +1 to −1 is used for vector creation.

A snippet of the following demonstrates the LSTM architecture with the outputs of model accuracy and model loss in the Figures 5.10 and 5.11, respectively.

```
import numpy as np
from sklearn.metrics import accuracy_score
```

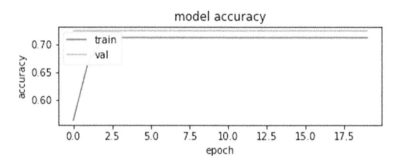

FIGURE 5.10
Model accuracy.

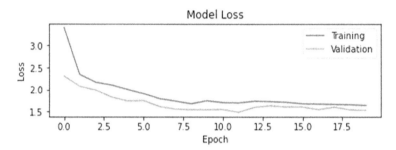

FIGURE 5.11
Model loss.

```
from tensorflow.keras.datasets import reuters
from tensorflow.keras.preprocessing.sequence import
pad_sequences
from tensorflow.keras.utils import to_categorical

num_words = 30000
max_length = 50
train_test_split = 0.3

(train_x, train_labels), (test_x, test_labels) = reuters.
    load_data(num_words = num_words, maxlen = max_length,
    test_split = train_test_split)

train_x = pad_sequences(train_x, padding = 'post')
test_x = pad_sequences(test_x, padding = 'post')

train_x = np.array(train_x).reshape((train_x.shape[0],
    train_x.shape[1], 1))
test_x = np.array(test_x).reshape((test_x.shape[0],
    test_x.shape[1], 1))
```

```
labels = np.concatenate((train_labels, test_labels))
labels = to_categorical(labels)

train_labels = labels[:1395]
test_labels = labels[1395:]

print(train_x.shape)
print(test_x.shape)
print(train_labels.shape)
print(test_labels.shape)

from tensorflow.keras.models import Sequential
from tensorflow.keras.layers import Dense, Activation
from tensorflow.keras import optimizers
from tensorflow.keras.wrappers.scikit_learn import
    KerasClassifier
from tensorflow.keras.layers import LSTM

def lstm():
    model = Sequential()
    model.add(LSTM(50, input_shape = (49,1), return_
        sequences = False))
    model.add(Dense(46))
    model.add(Dense(46))
    model.add(Activation('relu'))
    adam = optimizers.Adam(lr = 0.001)
    model.compile(loss = 'categorical_crossentropy',
        optimizer = adam, metrics = ['acc'])

    return model

model = KerasClassifier(build_fn = lstm, epochs = 20,
    batch_size = 50, verbose = 1)
history = model.fit(train_x, train_labels,
    validation_split=0.2)

Epoch 1/20
23/23 [==============================] - 3s 44ms/step -
    loss: 3.4020 - acc: 0.5636 - val_loss: 2.3086 - val_
    acc: 0.7240
Epoch 2/20
23/23 [==============================] - 1s 24ms/step -
    loss: 2.3468 - acc: 0.6703 - val_loss: 2.0781 - val_
    acc: 0.7240
Epoch 3/20
23/23 [==============================] - 1s 24ms/step -
    loss: 2.1666 - acc: 0.7124 - val_loss: 1.9911 - val_
    acc: 0.7240
```

```
Epoch 4/20
23/23 [==============================] - 1s 23ms/step -
   loss: 2.1018 - acc: 0.7124 - val_loss: 1.8292 - val_
   acc: 0.7240
Epoch 5/20
23/23 [==============================] - 1s 24ms/step -
   loss: 2.0046 - acc: 0.7124 - val_loss: 1.7428 - val_
   acc: 0.7240
Epoch 6/20
23/23 [==============================] - 1s 25ms/step -
   loss: 1.9120 - acc: 0.7124 - val_loss: 1.7529 - val_
   acc: 0.7240
Epoch 7/20
23/23 [==============================] - 1s 24ms/step -
   loss: 1.7969 - acc: 0.7124 - val_loss: 1.6178 - val_
   acc: 0.7240
Epoch 8/20
23/23 [==============================] - 1s 25ms/step -
   loss: 1.7400 - acc: 0.7124 - val_loss: 1.5582 - val_
   acc: 0.7240
Epoch 9/20
23/23 [==============================] - 1s 29ms/step -
   loss: 1.6802 - acc: 0.7124 - val_loss: 1.5450 - val_
   acc: 0.7240
Epoch 10/20
23/23 [==============================] - 1s 23ms/step -
   loss: 1.7475 - acc: 0.7124 - val_loss: 1.5430 - val_
   acc: 0.7240
Epoch 11/20
23/23 [==============================] - 1s 24ms/step -
   loss: 1.7067 - acc: 0.7124 - val_loss: 1.5501 - val_
   acc: 0.7240
Epoch 12/20
23/23 [==============================] - 1s 24ms/step -
   loss: 1.6996 - acc: 0.7124 - val_loss: 1.4881 - val_
   acc: 0.7240
Epoch 13/20
23/23 [==============================] - 1s 24ms/step -
   loss: 1.7402 - acc: 0.7124 - val_loss: 1.6013 - val_
   acc: 0.7240
Epoch 14/20
23/23 [==============================] - 1s 28ms/step -
   loss: 1.7299 - acc: 0.7124 - val_loss: 1.6354 - val_
   acc: 0.7240
Epoch 15/20
23/23 [==============================] - 1s 26ms/step -
   loss: 1.7138 - acc: 0.7124 - val_loss: 1.6082 - val_
   acc: 0.7240
```

```
Epoch 16/20
23/23 [==============================] - 1s 23ms/step -
   loss: 1.6810 - acc: 0.7124 - val_loss: 1.6085 - val_
   acc: 0.7240
Epoch 17/20
23/23 [==============================] - 1s 24ms/step -
   loss: 1.6739 - acc: 0.7124 - val_loss: 1.5489 - val_
   acc: 0.7240
Epoch 18/20
23/23 [==============================] - 1s 23ms/step -
   loss: 1.6664 - acc: 0.7124 - val_loss: 1.6056 - val_
   acc: 0.7240
Epoch 19/20
23/23 [==============================] - 1s 24ms/step -
   loss: 1.6591 - acc: 0.7124 - val_loss: 1.5425 - val_
   acc: 0.7240
Epoch 20/20
23/23 [==============================] - 1s 26ms/step -
   loss: 1.6436 - acc: 0.7124 - val_loss: 1.5348 - val_
   acc: 0.7240

pred_labels = model.predict(test_x)
test_labels_ = np.argmax(test_labels, axis = 1)
print("Test accuracy: "+str(accuracy_score(pred_labels,
   test_labels_)))

Test accuracy: 0.7479131886477463

from matplotlib import pyplot as plt

plt.subplot(211)
plt.plot(history.history['acc'])
plt.plot(history.history['val_acc'])
plt.title('model accuracy')
plt.ylabel('accuracy')
plt.xlabel('epoch')
plt.legend(['train', 'val'], loc='upper left')
plt.show()

plt.subplot(212)
plt.plot(history.history['loss'])
plt.plot(history.history['val_loss'])
plt.title('Model Loss')
plt.ylabel('Loss')
plt.xlabel('Epoch')
plt.legend(['Training', 'Validation'], loc='upper right')

plt.tight_layout()

plt.show()
```

5.1.6 Gated-Recurrent Units (GRU)

Gated-recurrent neural networks [5] are used for exploring the temporal and spatial data for applications related to speech recognition, machine translation and natural language processing. It is similar to LSTM with only two gates, namely the update gate and reset gate. The update gate is responsible for the flow of information into memory, and the reset gate controls the information that flows out of the memory. The main difference between LSTM and GRU is there is no memory unit that holds the state of the cell. A simple architecture of GRU is shown in Figure 5.12 and the parameters of the network are learned through Equations (5.13)–(5.16), where z represents the update gate and r represents the reset gate. The parameter t denotes the state of GRU at time t, and h_{t-1} is the hidden state value from the previous recurrent unit. The hidden layer h_t values are computed using the equations. These values are used for the further recurrent layers in the network.

$$z_t = \sigma(W_z.[h_{t-1}, x_t]) \tag{5.13}$$

$$r_t = \sigma(W_r.[h_{t-1}, x_t]) \tag{5.14}$$

$$\check{h}_t = \tan h(W.[r_t * h_{t-1}, x_t]) \tag{5.15}$$

$$h_t = (1 - z_t) * h_{t-1} + z_t * \check{h}_t \tag{5.16}$$

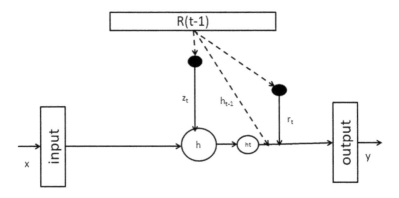

FIGURE 5.12
GRU architecture.The architecture of GPU with its components.

A snippet of the following demonstrates the GRU architecture with the out-puts of model accuracy and model loss in Figures 5.13 and 5.14, respectively.

```
import numpy as np

from sklearn.metrics import accuracy_score
from tensorflow.keras.datasets import reuters
from tensorflow.keras.preprocessing.sequence import
    pad_sequences
from tensorflow.keras.utils import to_categorical

num_words = 30000
max_length = 50
train_test_split = 0.3

(train_x, train_labels), (test_x, test_labels) = reuters.
    load_data(num_words = num_words, maxlen = max_length,
    test_split = train_test_split)
```

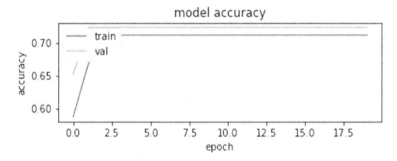

FIGURE 5.13
Model accuracy.

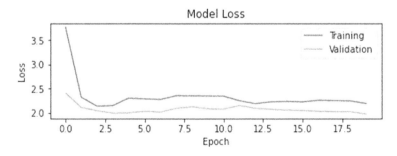

FIGURE 5.14
Model loss.

```
train_x = pad_sequences(train_x, padding = 'post')
test_x = pad_sequences(test_x, padding = 'post')

train_x = np.array(train_x).reshape((train_x.shape[0],
    train_x.shape[1], 1))
test_x = np.array(test_x).reshape((test_x.shape[0],
    test_x.shape[1], 1))

labels = np.concatenate((train_labels, test_labels))
labels = to_categorical(labels)

train_labels = labels[:1395]
test_labels = labels[1395:]

print(train_x.shape)
print(test_x.shape)
print(train_labels.shape)
print(test_labels.shape)

(1395, 49, 1)
(599, 49, 1)
(1395, 46)
(599, 46)

from tensorflow.keras.models import Sequential
from tensorflow.keras.layers import Dense, Activation
from tensorflow.keras import optimizers
from tensorflow.keras.wrappers.scikit_learn import
    KerasClassifier
from tensorflow.keras.layers import GRU

def gru():
    model = Sequential()
    model.add(GRU(50, input_shape = (49,1), return_
        sequences = False))
    model.add(Dense(46))
    model.add(Dense(46))
    model.add(Activation('relu'))
    adam = optimizers.Adam(lr = 0.001)
    model.compile(loss = 'categorical_crossentropy',
        optimizer = adam, metrics = ['acc'])

    return model

    model = KerasClassifier(build_fn = gru, epochs = 20,
        batch_size = 50, verbose = 1)
```

```
history = model.fit(train_x, train_labels,
   validation_split=0.2)

Epoch 1/20
23/23 [==============================] - 3s 50ms/step -
   loss: 3.7657 - acc: 0.5869 - val_loss: 2.4047 - val_
   acc: 0.6523
Epoch 2/20
23/23 [==============================] - 1s 30ms/step -
   loss: 2.3181 - acc: 0.6694 - val_loss: 2.1062 - val_
   acc: 0.7240
Epoch 3/20
23/23 [==============================] - 1s 30ms/step -
   loss: 2.1338 - acc: 0.7124 - val_loss: 2.0399 - val_
   acc: 0.7240
Epoch 4/20
23/23 [==============================] - 1s 29ms/step -
   loss: 2.1457 - acc: 0.7124 - val_loss: 1.9901 - val_
   acc: 0.7240
Epoch 5/20
23/23 [==============================] - 1s 30ms/step -
   loss: 2.2986 - acc: 0.7124 - val_loss: 1.9944 - val_
   acc: 0.7240
Epoch 6/20
23/23 [==============================] - 1s 31ms/step -
   loss: 2.2842 - acc: 0.7124 - val_loss: 2.0279 - val_
   acc: 0.7240
Epoch 7/20
23/23 [==============================] - 1s 29ms/step -
   loss: 2.2719 - acc: 0.7124 - val_loss: 2.0113 - val_
   acc: 0.7240
Epoch 8/20
23/23 [==============================] - 1s 30ms/step -
   loss: 2.3506 - acc: 0.7124 - val_loss: 2.0903 - val_
   acc: 0.7240
Epoch 9/20
23/23 [==============================] - 1s 30ms/step -
   loss: 2.3466 - acc: 0.7124 - val_loss: 2.1205 - val_
   acc: 0.7240
Epoch 10/20
23/23 [==============================] - 1s 30ms/step -
   loss: 2.3441 - acc: 0.7124 - val_loss: 2.0779 - val_
   acc: 0.7240
Epoch 11/20
23/23 [==============================] - 1s 31ms/step -
   loss: 2.3411 - acc: 0.7124 - val_loss: 2.0741 - val_
   acc: 0.7240
```

```
Epoch 12/20
23/23 [==============================] - 1s 30ms/step -
   loss: 2.2467 - acc: 0.7124 - val_loss: 2.1462 - val_
   acc: 0.7240
Epoch 13/20
23/23 [==============================] - 1s 30ms/step -
   loss: 2.1855 - acc: 0.7124 - val_loss: 2.0895 - val_
   acc: 0.7240
Epoch 14/20
23/23 [==============================] - 1s 30ms/step -
   loss: 2.2233 - acc: 0.7124 - val_loss: 2.0693 - val_
   acc: 0.7240
Epoch 15/20
23/23 [==============================] - 1s 30ms/step -
   loss: 2.2313 - acc: 0.7124 - val_loss: 2.0514 - val_
   acc: 0.7240
Epoch 16/20
23/23 [==============================] - 1s 32ms/step -
   loss: 2.2224 - acc: 0.7124 - val_loss: 2.0427 - val_
   acc: 0.7240
Epoch 17/20
23/23 [==============================] - 1s 30ms/step -
   loss: 2.2562 - acc: 0.7124 - val_loss: 2.0281 - val_
   acc: 0.7240
Epoch 18/20
23/23 [==============================] - 1s 30ms/step -
   loss: 2.2488 - acc: 0.7124 - val_loss: 2.0217 - val_
   acc: 0.7240
Epoch 19/20
23/23 [==============================] - 1s 30ms/step -
   loss: 2.2426 - acc: 0.7124 - val_loss: 2.0184 - val_
   acc: 0.7240
Epoch 20/20
23/23 [==============================] - 1s 31ms/step -
   loss: 2.1889 - acc: 0.7124 - val_loss: 1.9713 - val_
   acc: 0.7240

pred_labels = model.predict(test_x)
test_labels_ = np.argmax(test_labels, axis = 1)
print("Test accuracy: "+str(accuracy_score(pred_labels,
   test_labels_)))

Test accuracy: 0.7479131886477463
from matplotlib import pyplot as plt

plt.subplot(211)
plt.plot(history.history['acc'])
plt.plot(history.history['val_acc'])
plt.title('model accuracy')
```

```
plt.ylabel('accuracy')
plt.xlabel('epoch')
plt.legend(['train', 'val'], loc='upper left')
plt.show()

plt.subplot(212)
plt.plot(history.history['loss'])
plt.plot(history.history['val_loss'])
plt.title('Model Loss')
plt.ylabel('Loss')
plt.xlabel('Epoch')
plt.legend(['Training', 'Validation'], loc='upper right')

plt.tight_layout()

plt.show()
```

References

1. Li, Z., Liu, F., Yang, W., Peng, S., & Zhou, J. (2021). A survey of convolutional neural networks: analysis, applications, and prospects. *IEEE Transactions on Neural Networks and Learning Systems*, 33(12), 6999–7019.
2. O'Shea, K., & Nash, R. (2015). An introduction to convolutional neural networks. arXiv preprint arXiv:1511.08458.
3. Koushik, J. (2016). Understanding convolutional neural networks. arXiv preprint arXiv:1605.09081.
4. Yu, Y., Si, X., Hu, C., & Zhang, J. (2019). A review of recurrent neural networks: LSTM cells and network architectures. *Neural Computation*, 31(7), 1235–1270.
5. Sachin, S., Tripathi, A., Mahajan, N., Aggarwal, S., & Nagrath, P. (2020). Sentiment analysis using gated recurrent neural networks. *SN Computer Science*, 1(2), 1–13.

Exercises

1. Given an image of size $n \times n$, filter f $\times$ f, stride s and padding p, determine the formula for size of the feature map. Assuming an input image of size 15 $\times$ 15, with a filter of size 4 $\times$ 4 and stride 2, determine the size of the convolved matrix.

2. What are the two biggest issues that arise from the convolution layer in a CNN, and how can they be resolved?

3. What happens to the gradient if you backpropagate through a long sequence in an RNN? How can you overcome this problem?

4. What are the major differences in the operation of an LTSM and a GRU?

5. Given below is the code to create a dataset where the input is the summation of three sinusoids.

```
num_train_data = 4000
num_test_data = 1000
timestep = 0.1
tm =  np.arange(0, (num_train_data+num_test_
    data)*timestep, timestep);
y = np.sin(tm) + np.sin(tm*np.pi/2) + np.sin(tm*(-3*np.
    pi/2))
```

Using this dataset, create a simple CNN and LSTM and compare their mean squared errors. What conclusions can be drawn from the performance of these models?

6

Cloud Examples for Neural Networks Multi-Modal Architectures

6.1 AlexNet

Convolutional neural networks (CNNs) are the most powerful tools for computer vision applications. AlexNet is one such CNN that can be used for relatively large datasets and wide areas as well [1]. In this section, we discuss the architecture and implementation of AlexNet using Keras as the backend. AlexNet is an eight-layer CNN, as shown in Figure 6.1. It was a part of the ImageNet challenge. The baseline of the network is tested on the ImageNet dataset. Here, for implementation purposes, CIFAR dataset is used. The architecture and the implementations are discussed as follows.

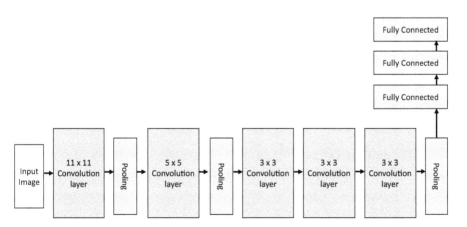

FIGURE 6.1
AlexNet architecture.

DOI: 10.1201/9781003215974-8

6.1.1 Architecture

The total number of convolution layers present in the AlexNet is eight, including the convolutions and pooling. The first convolution layer begins with the shape 11 × 11 with stride 4. The first layer shape is more as the image has to be captured initially. It is followed by a max-pooling layer with stride 2. The second convolution layer shape is 5 × 5 with padding 2. Here, the padding is introduced for capturing the corners of the image accurately. It is then followed by the pooling layer of size 3 × 3 with stride 2. It is then followed by three convolution layers of size 3 × 3. The final layer of max-pooling is considered with the size 3 × 3 and stride 2. The last layers of fully connected are used for the classification.

The activation functions that are used as a part of the AlexNet is ReLU activation function. It makes the gradient of the parameters close to 1, as compared to the sigmoid activation function. The fully connected layers in the AlexNet consist of 4,096 outputs. These layers make the AlexNet model size of approximately 1 GB. Hence, sufficient amount of GPU is needed for the training. We use the Colab as the infrastructure that provides sufficient GPU for training and testing of the model.

6.1.2 Implementation

In this section, we show the implementation of the AlexNet model using Keras as the backend. We describe in detail the dataset, pre-processing, training and visualization using CIFAR-10 as the dataset [2].

6.1.3 Imports

The necessary libraries that are needed for imports are numpy, tensorflow, train_test_split and others are included, as shown in the code snippet below.

```
import numpy as np
import matplotlib.pyplot as plt
import tensorflow as tf
from random import randint
from sklearn.model_selection import train_test_split
```

6.1.4 Reading the Dataset

The dataset that is considered for the implementation is CIFAR-10. It consists of 60,000 images of shape 32 × 32 × 3 with ten classes. The training set consists of 50,000 images and test set consists of 10,000 images. The classes that are present in the dataset are [0 – airplane, 1-automobile, 2-bird, 3-cat, 4-deer, 5-dog, 6-frog, 7-horse, 8-ship and 9-truck].

The code snippet for splitting the dataset into training and testing is shown below. The keras.datasets() are used for loading the dataset, and the number of training images and the testing images are seen using the shape() function. The different classes that are present in the dataset are also visualized using the matplotlib function as shown in Figure 6.2.

```
(train_images,train_lables),(test_images,test_lables) =
    tf.keras.datasets.cifar10.load_data()
no_of_train_images = train_images.shape[0]
image_shape = train_images.shape[1:]

print("No of training images =",no_of_train_images,"each
    of shape =",image_shape)
print("No of     test images =",test_images.
    shape[0],"each of shape =",test_images.shape[1:])

No of training images = 50000 each of shape = (32, 32, 3)
No of test images = 10000 each of shape = (32, 32, 3)

classes = list("airplane,automobile,bird,cat,deer,dog,fro
    g,horse,ship,truck".split(","))
print(classes)

['airplane', 'automobile', 'bird', 'cat', 'deer', 'dog',
    'frog', 'horse', 'ship', 'truck']
```

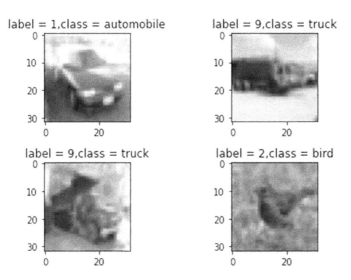

FIGURE 6.2
Classes in CIFAR-10 dataset.

```
for i in range(4):
    k = randint(0,no_of_train_images)
    plt.subplot(2,2,i+1)
    plt.imshow(train_images[k])
    plt.title("label = "+str(train_lables[k][0])+",class
        = "+classes[train_lables[k][0]])
plt.tight_layout()
```

6.1.5 One-Hot Encoding

The labels of the images are in the text form as 'truck', bird', etc. These labels have to be converted into categories using the one-hot encoding scheme. The one-hot encoding refers to the conversion of the labels into categories as 1, 2, 3, etc., depending on the data. Here, the CIFAR-10 dataset consists of ten classes, and hence the number of categories for classification should be ten. The code snippet for the conversion into categorical values is shown below.

The function to_categorical() is used for the conversion of the labels into different categories. Some examples of visualizations are shown as the output using the matplotlib function(). The first image has an encoding of [0,0,0,0,0,0,1,0,0,0]. The index with the value of 1 is 6. So this image corresponds to label 6, which is a frog shown in Figure 6.3. The second and third images have an encoding of [0,0,0,0,0,0,0,0,0,1]. The index with the value of 1 is 9. So this image corresponds to label 9, which is a truck, as shown in Figures 6.4 and 6.5. Similarly, the fourth image has an encoding of [0,0,0,0,1,0,0,0,0,0]. The index with the value of 1 is 5. So this image corresponds to label 5, which is a deer shown in Figure 6.6. Likewise, the other image has an encoding of [0,1,0,0,0,0,0,0,0,0]. The index with the value of 1 is 2. So this image corresponds to label 2, which is an automobile, as shown in Figure 6.7.

```
train_lables = tf.keras.utils.
to_categorical(train_lables,num_classes=10)
test_lables = tf.keras.utils.
to_categorical(test_lables,num_classes=10)

plt.imshow(train_images[0])
one_hot = train_lables[0]
lable = np.argmax(one_hot)
plt.title(f"One hot encoding = {one_hot}, label =
{lable}, class = {classes[lable]}")

Text(0.5, 1.0, 'One hot encoding = [0. 0. 0. 0. 0. 0. 1.
0. 0. 0.], label = 6, class = frog')
```

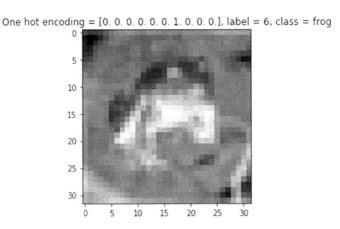

FIGURE 6.3
One-hot encoding of class-frog.

```
plt.imshow(train_images[1])
one_hot = train_lables[1]
lable = np.argmax(one_hot)
plt.title(f"One hot encoding = {one_hot}, label =
    {lable}, class = {classes[lable]}")

Text(0.5, 1.0, 'One hot encoding = [0. 0. 0. 0. 0. 0. 0.
    0. 0. 1.], label = 9, class = truck')
```

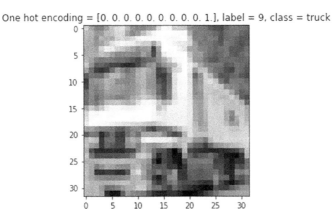

FIGURE 6.4
One-hot encoding of class-truck.

```
plt.imshow(train_images[2])
one_hot = train_lables[2]
```

```
lable = np.argmax(one_hot)
plt.title(f"One hot encoding = {one_hot}, label =
    {lable}, class = {classes[lable]}")
Text(0.5, 1.0, 'One hot encoding = [0. 0. 0. 0. 0. 0. 0.
    0. 0. 1.], label = 9, class = truck')
```

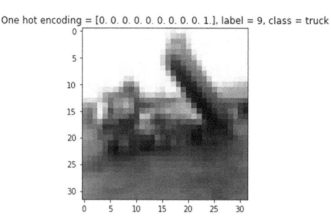

FIGURE 6.5
One-hot encoding of class-truck.

```
plt.imshow(train_images[3])
one_hot = train_lables[3]
lable = np.argmax(one_hot)
plt.title(f"One hot encoding = {one_hot}, label =
    {lable}, class = {classes[lable]}")
Text(0.5, 1.0, 'One hot encoding = [0. 0. 0. 0. 1. 0. 0.
    0. 0. 0.], label = 4, class = deer')
```

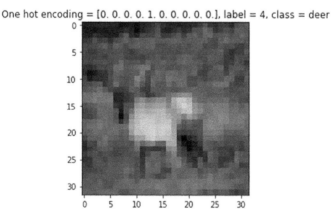

FIGURE 6.6
One-hot encoding of class-deer.

```
plt.imshow(train_images[4])
one_hot = train_lables[4]
lable = np.argmax(one_hot)
plt.title(f"One hot encoding = {one_hot}, label =
   {lable}, class = {classes[lable]}")
Text(0.5, 1.0, 'One hot encoding = [0. 1. 0. 0. 0. 0. 0.
   0. 0. 0.], label = 1, class = automobile')
```

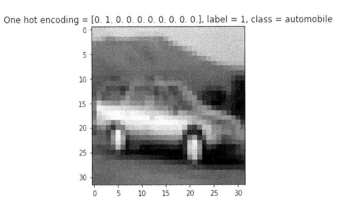

FIGURE 6.7
One-hot encoding of class-automobile.

6.1.6 Data Augmentation

Once the one-hot encoding is completed, the training and testing images are obtained using the function train_test_split() with the test size as 0.1. The image sizes required for the AlexNet are augmented first using the ImageDataGenerator() function with rescaling and rotation. Now, the data-set is ready for training using the AlexNet architecture. The implementation details of the AlexNet are discussed in the next sections.

```
train_images, val_images, train_lables, val_lables
= train_test_split(train_images, train_lables,
test_size=0.1)

train_datagen = tf.keras.preprocessing.image.ImageDataGen
   erator(rescale=1./255, width_shift_range=0.1, height_
   shift_range=0.1, zoom_range=0.1, shear_range=0.1,
   rotation_range=2)
train_datagen.fit(train_images)
training_set = train_datagen.flow(train_images, train_
   lables, batch_size=32)
```

```
val_datagen = tf.keras.preprocessing.image.ImageDataGener
    ator(rescale=1./255)
val_datagen.fit(val_images)
val_set = val_datagen.flow(val_images, val_lables,
    batch_size=32)
```

6.1.6.1 Training using AlextNet

The AlexNet implementation is shown in Figure 6.8 and the code snippet is shown below.

```
alexnet_model = tf.keras.models.Sequential(
    [
        tf.keras.layers.Conv2D(filters=32,kernel_
            size=3,activation="relu",input_shape=image_
            shape),
        tf.keras.layers.
            MaxPool2D(pool_size=(2,2),strides=2),
        tf.keras.layers.Conv2D(filters=32,kernel_size=3,pad
            ding="same",activation="relu"),
```

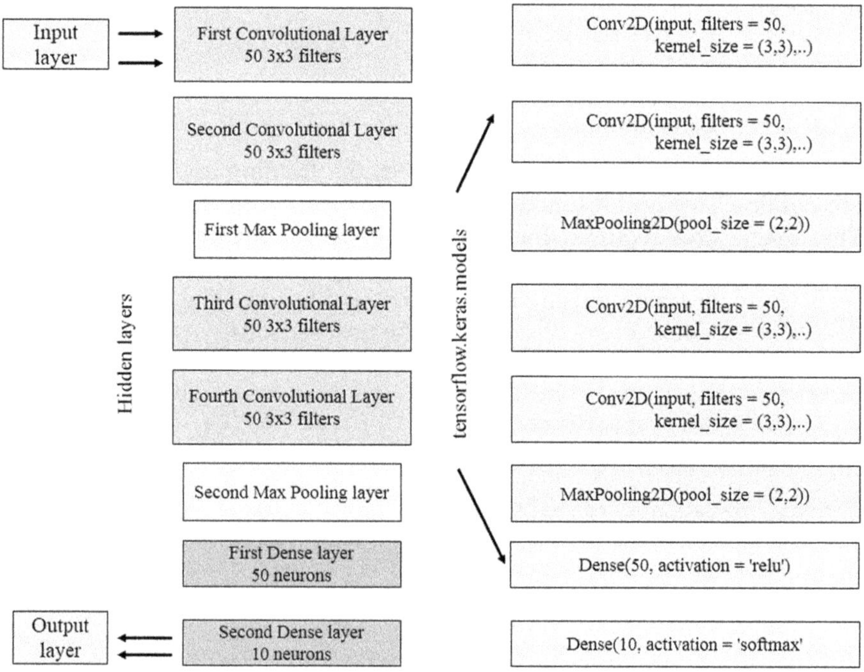

FIGURE 6.8
AlexNet implementation.

```
        tf.keras.layers.
          MaxPool2D(pool_size=(2,2),strides=2),
        tf.keras.layers.Conv2D(filters=64,kernel_size=3,pad
          ding="same",activation="relu"),
        tf.keras.layers.Conv2D(filters=64,kernel_size=3,pad
          ding="same",activation="relu"),
        tf.keras.layers.Conv2D(filters=64,kernel_size=3,pad
          ding="same",activation="relu"),
        tf.keras.layers.
          MaxPool2D(pool_size=(2,2),strides=2),
        tf.keras.layers.Flatten(),
        tf.keras.layers.Dense(units=4096,activation="r
          elu"),
        tf.keras.layers.Dropout(rate=0.1),
        tf.keras.layers.Dense(units=4096,activation="r
          elu"),
        tf.keras.layers.Dropout(rate=0.1),
        tf.keras.layers.Dense(10,activation="softmax")
    ]
    )

alexnet_model.summary()
Model: "sequential_5"
```

Layer (type)	Output Shape	Param #
conv2d_25 (Conv2D)	(None, 30, 30, 32)	896
max_pooling2d_15 (MaxPooling	(None, 15, 15, 32)	0
conv2d_26 (Conv2D)	(None, 15, 15, 32)	9248
max_pooling2d_16 (MaxPooling	(None, 7, 7, 32)	0
conv2d_27 (Conv2D)	(None, 7, 7, 64)	18496
conv2d_28 (Conv2D)	(None, 7, 7, 64)	36928
conv2d_29 (Conv2D)	(None, 7, 7, 64)	36928
max_pooling2d_17 (MaxPooling	(None, 3, 3, 64)	0
flatten_5 (Flatten)	(None, 576)	0
dense_15 (Dense)	(None, 4096)	2363392
dropout_8 (Dropout)	(None, 4096)	0

```
dense_16 (Dense)                  (None, 4096)                16781312
```

```
dropout_9 (Dropout)               (None, 4096)                    0
```

```
dense_17 (Dense)                  (None, 10)                    40970
==============================================================
```

```
Total params: 19,288,170
Trainable params: 19,288,170
Non-trainable params: 0
```

```
alexnet_model.compile(optimizer=tf.keras.optimizers.
    Adam(learning_rate=0.001), loss='categorical_
    crossentropy', metrics=['accuracy'])
```

6.1.6.2 Validation using AlextNet

The Alexnet model is trained using the training set and validated against the validation set for 20 epochs with 32 images trained per batch. But the training will be stopped mid-way if there is no increase in validation accuracy for two consecutive epochs

```
training_history = alexnet_model.fit
                   (training_set,
                   batch_size=32,
                   epochs=20,
                   validation_data = val_set,
                   callbacks = [tf.keras.callbacks.
                       EarlyStopping(monitor='val_
                       accuracy',patience=2,restore_best_
                       weights=True)]
                   )
```

```
Epoch 1/20
1407/1407 [==============================] - 51s 36ms/
    step - loss: 1.6985 - accuracy: 0.3585 - val_loss:
    1.3692 - val_accuracy: 0.5066
Epoch 2/20
1407/1407 [==============================] - 50s 35ms/
    step - loss: 1.3542 - accuracy: 0.5055 - val_loss:
    1.2019 - val_accuracy: 0.5632
Epoch 3/20
1407/1407 [==============================] - 50s 36ms/
    step - loss: 1.1982 - accuracy: 0.5715 - val_loss:
    1.0943 - val_accuracy: 0.6114
Epoch 4/20
1407/1407 [==============================] - 51s 36ms/
    step - loss: 1.0994 - accuracy: 0.6090 - val_loss:
    1.0958 - val_accuracy: 0.6034
```

```
Epoch 5/20
1407/1407 [==============================] - 51s 36ms/
    step - loss: 1.0286 - accuracy: 0.6359 - val_loss:
    0.9791 - val_accuracy: 0.6636
Epoch 6/20
1407/1407 [==============================] - 50s 36ms/
    step - loss: 0.9763 - accuracy: 0.6564 - val_loss:
    0.9620 - val_accuracy: 0.6608
Epoch 7/20
1407/1407 [==============================] - 51s 36ms/
    step - loss: 0.9312 - accuracy: 0.6730 - val_loss:
    0.9008 - val_accuracy: 0.6882
Epoch 8/20
1407/1407 [==============================] - 50s 36ms/
    step - loss: 0.8930 - accuracy: 0.6837 - val_loss:
    0.8693 - val_accuracy: 0.6940
Epoch 9/20
1407/1407 [==============================] - 50s 36ms/
    step - loss: 0.8696 - accuracy: 0.6928 - val_loss:
    0.8638 - val_accuracy: 0.6978
Epoch 10/20
1407/1407 [==============================] - 50s 36ms/
    step - loss: 0.8477 - accuracy: 0.7030 - val_loss:
    0.8559 - val_accuracy: 0.7092
Epoch 11/20
1407/1407 [==============================] - 50s 36ms/
    step - loss: 0.8225 - accuracy: 0.7108 - val_loss:
    0.8199 - val_accuracy: 0.7210
Epoch 12/20
1407/1407 [==============================] - 50s 36ms/
    step - loss: 0.8090 - accuracy: 0.7155 - val_loss:
    0.8788 - val_accuracy: 0.7116
Epoch 13/20
1407/1407 [==============================] - 51s 36ms/
    step - loss: 0.7901 - accuracy: 0.7263 - val_loss:
    0.8431 - val_accuracy: 0.7160
```

6.1.6.3 Accuracy and Loss Estimation with AlexNet

The accuracy and the loss of the implemented AlexNet model are visualized using the matplot library. The function plot() is used with the training and validation for accuracy. The accuracy plot is shown in Figure 6.9. It can be seen that the training accuracy of the model is 72.63% and validation accuracy is 71.60%. Hence, the validation is on par with the training. Similarly, the loss of the model is shown in Figure 6.10. It can be seen that the validation loss is less compared to the training loss. Therefore, the model has met the requirements for the generalization of the model.

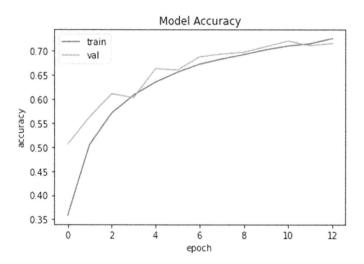

FIGURE 6.9
Accuracy of AlexNet model.

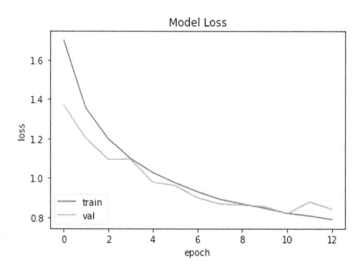

FIGURE 6.10
Loss of AlexNet model.

```
plt.plot(training_history.history["accuracy"],label="train")
plt.plot(training_history.
history["val_accuracy"],label="val")
plt.title("Model Accuracy")
plt.ylabel('accuracy')
plt.xlabel('epoch')
plt.legend(loc="upper left")
```

```
plt.plot(training_history.history["loss"],label="train")
plt.plot(training_history.history["val_loss"],label="val")
plt.title("Model Loss")
plt.ylabel('loss')
plt.xlabel('epoch')
plt.legend(loc="lower left")
```

6.2 VGG-16

In the previous section, we have seen the implementation of AlexNet with the CIFAR-10 dataset. It did not provide much accuracy for the considered dataset. In this section, we describe an architecture VGG-16 network that has more convolution layers than the AlexNet. In this section, the detailed architecture with implementation is discussed.

6.2.1 Architecture

The architecture of VGG-16 consists of 16 convolutional layers as shown in Figure 6.11 [3]. The input to the network is of the size 224 × 224. The first two layers have 64 filters, which is followed by another two layers with 128 filters. The next layers are with 256 filters and 512 filters. It is then followed by the fully connected layers for the classifications. The summary of the layers and components are shown in Table 6.1.

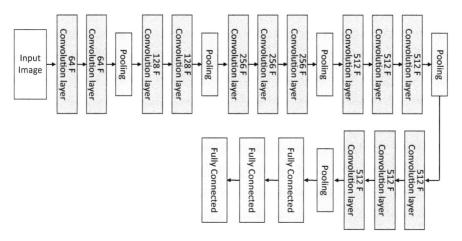

FIGURE 6.11
VGG-16 architecture.

TABLE 6.1

VGG-16 Architecture and Components

Number	Convolution
Layer 1,2	Convolution layer of 64 channel of 3 × 3 kernel with padding 1 and stride 1
Layer 3,4	Convolution layer of 128 channel of 3 × 3 kernel
Layer 5,6,7	Convolution layer of 256 channel of 3 × 3 kernel
Layer 8,9,10	Convolution layer of 512 channel of 3 × 3 kernel
Layer 11,12,13	Convolution layer of 512 channel of 3 × 3 kernel

6.2.2 Implementation

In this section, we show the implementation of the VGG-16 model using Keras as the backend. We describe in detail the dataset, pre-processing, training and visualization using CIFAR-10 as the dataset.

6.2.3 Imports

The import libraries used are numpy, mtplotlib, tensorflow and model_selection for the implementation of the VGG-16 network.

```
import numpy as np
import matplotlib.pyplot as plt
import tensorflow as tf
from random import randint
from sklearn.model_selection import train_test_split
```

6.2.4 Reading the Dataset

The dataset that is considered for the implementation is CIFAR-10 [2]. It consists of 60,000 images of shape 32 × 32 × 3 with ten classes. The training set consists of 50,000 images and test set consists of 10,000 images. The classes that are present in the dataset are [0 – airplane, 1-automobile, 2-bird, 3-cat, 4-deer, 5-dog, 6-frog, 7-horse, 8-ship and 9-truck].

   The code snippet for splitting the dataset into training and testing is shown below. The keras.datasets() is used for loading the dataset and the number of training images and the testing images are seen using the shape() function. The different classes that are present in the dataset are also visualized using the matplotlib function as shown in Figure 6.12.

```
(train_images,train_lables1),(test_images,test_lables1) =
    tf.keras.datasets.cifar10.load_data()
no_of_train_images = train_images.shape[0]
image_shape = train_images.shape[1:]
```

label = 7,class = horse

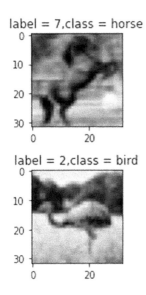

label = 5,class = dog

label = 2,class = bird

label = 9,class = truck

FIGURE 6.12
CIFAR-10 dataset for VGG-16.

```
print("No of training images =",no_of_train_images,"each
    of shape =",image_shape)
print("No of     test images =",test_images.
    shape[0],"each of shape =",test_images.shape[1:])

No of training images = 50000 each of shape = (32, 32, 3)
No of test images = 10000 each of shape = (32, 32, 3)

classes = list("airplane,automobile,bird,cat,deer,dog,
    frog,horse,ship,truck".split(","))
print(classes)

['airplane', 'automobile', 'bird', 'cat', 'deer', 'dog',
    'frog', 'horse', 'ship', 'truck']

for i in range(4):
  k = randint(0,no_of_train_images)
  plt.subplot(2,2,i+1)
  plt.imshow(train_images[k])
  plt.title("label = "+str(train_lables1[k][0])+",class =
    "+classes[train_lables1[k][0]])
plt.tight_layout()
```

6.2.5 One-Hot Encoding

The labels of the images are in the text form as 'truck', bird', etc. These labels have to be converted into categories using the one-hot encoding scheme. The one-hot encoding refers to the conversion of the labels into categories as 1, 2, 3, etc., depending on the data. Here, the CIFAR-10 dataset consists of ten classes, and hence the number of categories for classification should be ten. The encoding implementation is the same as the AlexNet, as discussed in the previous section. The code snippet for the conversion into categorical values is shown below. The function to_categorical() is used for the conversion of the labels into different categories. Some of the example visualizations are shown as the output using the matplotlib function(). The first image has an encoding of [0, 0, 0, 0, 0, 0, 1, 0, 0, 0]. The index where the value is 1 is 6. So this image corresponds to label 6, which is a frog as shown in Figure 6.13.

```
train_lables = tf.keras.utils.to_categorical
    (train_lables,num_classes=10)
test_lables = tf.keras.utils.
    to_categorical(test_lables,num_classes=10)

plt.imshow(train_images[0])
one_hot = train_lables[0]
lable = np.argmax(one_hot)
plt.title(f"One hot encoding = {one_hot}, label =
    {lable}, class = {classes[lable]}")

Text(0.5, 1.0, 'One hot encoding = [0. 0. 0. 0. 0. 0. 1.
    0. 0. 0.], label = 6, class = frog')
```

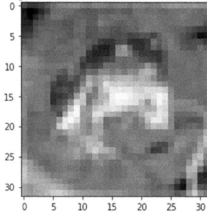

One hot encoding = [0. 0. 0. 0. 0. 0. 1. 0. 0. 0.], label = 6, class = frog

FIGURE 6.13
One-hot encoding of class-frog in VGG-16.

6.2.6 Data Augmentation

Once the one-hot encoding is completed, the training and testing images are obtained using the function train_test_split() with a test size of 0.1. The image sizes required for the AlexNet are augmented first using the ImageDataGenerator() function with rescaling and rotation. Now, the dataset is ready for training using the AlexNet architecture. The implementation details of the VGG-16 are discussed in the next sections.

```
train_datagen = tf.keras.preprocessing.image.ImageDataGen
    erator(rescale=1./255, width_shift_range=0.1, height_
    shift_range=0.1, zoom_range=0.1, shear_range=0.1,
    rotation_range=2)
train_datagen.fit(train_images)
training_set = train_datagen.flow(train_images, train_
    lables, batch_size=32)

val_datagen = tf.keras.preprocessing.image.ImageDataGener
    ator(rescale=1./255)
val_datagen.fit(val_images)
val_set = val_datagen.flow(val_images, val_lables,
    batch_size=32)
```

6.2.6.1 Training using VGG-16

The VGG-16 implementation is shown in Figure 6.14 and the code snippet as below.

```
vgg16_model = tf.keras.models.Sequential(
    [
        tf.keras.layers.Conv2D(filters=64,kernel_siz
            e=3,activation="relu",padding="same",in
            put_shape=image_shape),
        tf.keras.layers.Conv2D(filters=64,kernel_size=3,pad
            ding="same",activation="relu"),
        tf.keras.layers.MaxPool2D
            (pool_size=(2,2),strides=2),
        tf.keras.layers.BatchNormalization(),
        tf.keras.layers.Conv2D(filters=128,kernel_size=3,pa
            dding="same",activation="relu"),
        tf.keras.layers.Conv2D(filters=128,kernel_size=3,pa
            dding="same",activation="relu"),
        tf.keras.layers.
            MaxPool2D(pool_size=(2,2),strides=2),
        tf.keras.layers.BatchNormalization(),
        tf.keras.layers.Conv2D(filters=256,kernel_size=3,pa
            dding="same",activation="relu"),
```

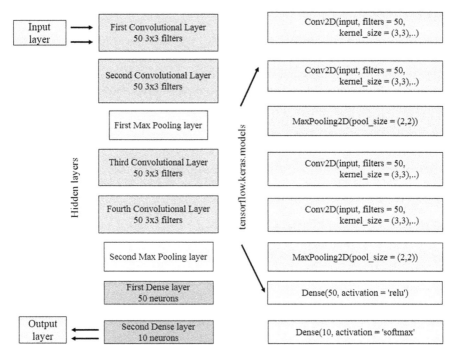

FIGURE 6.14
VGG-16 implementation.

```
tf.keras.layers.Conv2D(filters=256,kernel_size=3,pa
    dding="same",activation="relu"),
tf.keras.layers.Conv2D(filters=256,kernel_size=3,pa
    dding="same",activation="relu"),
tf.keras.layers.
    MaxPool2D(pool_size=(2,2),strides=2),
tf.keras.layers.BatchNormalization(),
tf.keras.layers.Conv2D(filters=512,kernel_size=3,pa
    dding="same",activation="relu"),
tf.keras.layers.Conv2D(filters=512,kernel_size=3,pa
    dding="same",activation="relu"),
tf.keras.layers.Conv2D(filters=512,kernel_size=3,pa
    dding="same",activation="relu"),
tf.keras.layers.MaxPool2D
    (pool_size=(2,2),strides=2),
tf.keras.layers.BatchNormalization(),
tf.keras.layers.Conv2D(filters=512,kernel_size=3,pa
    dding="same",activation="relu"),
tf.keras.layers.Conv2D(filters=512,kernel_size=3,pa
    dding="same",activation="relu"),
```

```
        tf.keras.layers.Conv2D(filters=512,kernel_size=3,pa
            dding="same",activation="relu"),
        tf.keras.layers.
            MaxPool2D(pool_size=(2,2),strides=2),
        tf.keras.layers.BatchNormalization(),
        tf.keras.layers.Flatten(),
        tf.keras.layers.Dense(units=4096,activation="relu"),
        tf.keras.layers.Dropout(rate=0.3),
        tf.keras.layers.Dense(units=4096,activation="r
            elu"),
        tf.keras.layers.Dropout(rate=0.3),
        tf.keras.layers.Dense(10,activation="softmax")
    ]
    )

vgg16_model.summary()

Model: "sequential_3"
```

Layer (type)	Output Shape	Param #
conv2d_39 (Conv2D)	(None, 32, 32, 64)	1792
conv2d_40 (Conv2D)	(None, 32, 32, 64)	36928
max_pooling2d_15 (MaxPooling	(None, 16, 16, 64)	0
batch_normalization_10 (Batc	(None, 16, 16, 64)	256
conv2d_41 (Conv2D)	(None, 16, 16, 128)	73856
conv2d_42 (Conv2D)	(None, 16, 16, 128)	147584
max_pooling2d_16 (MaxPooling	(None, 8, 8, 128)	0
batch_normalization_11 (Batc	(None, 8, 8, 128)	512
conv2d_43 (Conv2D)	(None, 8, 8, 256)	295168
conv2d_44 (Conv2D)	(None, 8, 8, 256)	590080
conv2d_45 (Conv2D)	(None, 8, 8, 256)	590080
max_pooling2d_17 (MaxPooling	(None, 4, 4, 256)	0
batch_normalization_12 (Batc	(None, 4, 4, 256)	1024

conv2d_46 (Conv2D)	(None, 4, 4, 512)	1180160
conv2d_47 (Conv2D)	(None, 4, 4, 512)	2359808
conv2d_48 (Conv2D)	(None, 4, 4, 512)	2359808
max_pooling2d_18 (MaxPooling	(None, 2, 2, 512)	0
batch_normalization_13 (Batc	(None, 2, 2, 512)	2048
conv2d_49 (Conv2D)	(None, 2, 2, 512)	2359808
conv2d_50 (Conv2D)	(None, 2, 2, 512)	2359808
conv2d_51 (Conv2D)	(None, 2, 2, 512)	2359808
max_pooling2d_19 (MaxPooling	(None, 1, 1, 512)	0
batch_normalization_14 (Batc	(None, 1, 1, 512)	2048
flatten_3 (Flatten)	(None, 512)	0
dense_9 (Dense)	(None, 4096)	2101248
dropout_16 (Dropout)	(None, 4096)	0
dense_10 (Dense)	(None, 4096)	16781312
dropout_17 (Dropout)	(None, 4096)	0
dense_11 (Dense)	(None, 10)	40970

```
=============================================================
Total params: 33,644,106
Trainable params: 33,641,162
Non-trainable params: 2,944
```

6.2.6.2 Validation using VGG-16

The VGG-16 model is trained using the training set and validated against the validation set for 20 epochs with 32 images trained per batch. But the training will be stopped mid-way if there is no increase in validation accuracy for two consecutive epochs.

```
vgg16_model.compile(optimizer=tf.keras.optimizers.
    Adam(learning_rate=0.001), loss='categorical_
    crossentropy', metrics=['accuracy'])
```

```
training_history = vgg16_model.fit
                    (training_set,
                    batch_size=32,
                    epochs=20,
                    validation_data = val_set,
                    callbacks = [tf.keras.callbacks.
                        EarlyStopping(monitor='val_
                        accuracy',patience=2,restore_
                        best_weights=True)]
                    )
Epoch 1/20
1407/1407 [==============================] - 102s 72ms/
    step - loss: 1.9631 - accuracy: 0.2472 - val_loss:
    1.8535 - val_accuracy: 0.3062
Epoch 2/20
1407/1407 [==============================] - 100s 71ms/
    step - loss: 1.6896 - accuracy: 0.3378 - val_loss:
    1.5638 - val_accuracy: 0.3978
Epoch 3/20
1407/1407 [==============================] - 101s 71ms/
    step - loss: 1.5071 - accuracy: 0.4257 - val_loss:
    1.3002 - val_accuracy: 0.5060
Epoch 4/20
1407/1407 [==============================] - 101s 71ms/
    step - loss: 1.2827 - accuracy: 0.5421 - val_loss:
    1.3099 - val_accuracy: 0.5436
Epoch 5/20
1407/1407 [==============================] - 101s 71ms/
    step - loss: 1.1220 - accuracy: 0.6143 - val_loss:
    1.1175 - val_accuracy: 0.6294
Epoch 6/20
1407/1407 [==============================] - 101s 71ms/
    step - loss: 0.9894 - accuracy: 0.6685 - val_loss:
    0.8676 - val_accuracy: 0.7016
Epoch 7/20
1407/1407 [==============================] - 101s 72ms/
    step - loss: 0.8746 - accuracy: 0.7104 - val_loss:
    1.0312 - val_accuracy: 0.6768
Epoch 8/20
1407/1407 [==============================] - 101s 72ms/
    step - loss: 0.8213 - accuracy: 0.7279 - val_loss:
    0.8857 - val_accuracy: 0.7140
Epoch 9/20
1407/1407 [==============================] - 100s 71ms/
    step - loss: 0.7108 - accuracy: 0.7691 - val_loss:
    0.7449 - val_accuracy: 0.7478
```

```
Epoch 10/20
1407/1407 [==============================] - 101s 72ms/
    step - loss: 0.6507 - accuracy: 0.7883 - val_loss:
    0.6522 - val_accuracy: 0.7858
Epoch 11/20
1407/1407 [==============================] - 101s 72ms/
    step - loss: 0.6193 - accuracy: 0.8013 - val_loss:
    0.8387 - val_accuracy: 0.7318
Epoch 12/20
1407/1407 [==============================] - 101s 72ms/
    step - loss: 0.5533 - accuracy: 0.8205 - val_loss:
    0.5994 - val_accuracy: 0.8006
Epoch 13/20
1407/1407 [==============================] - 101s 72ms/
    step - loss: 0.5046 - accuracy: 0.8379 - val_loss:
    0.5605 - val_accuracy: 0.8148
Epoch 14/20
1407/1407 [==============================] - 101s 72ms/
    step - loss: 0.4716 - accuracy: 0.8490 - val_loss:
    0.6169 - val_accuracy: 0.8046
Epoch 15/20
1407/1407 [==============================] - 101s 72ms/
    step - loss: 0.4415 - accuracy: 0.8592 - val_loss:
    0.5339 - val_accuracy: 0.8272
Epoch 16/20
1407/1407 [==============================] - 101s 71ms/
    step - loss: 0.4062 - accuracy: 0.8714 - val_loss:
    0.6467 - val_accuracy: 0.7986
Epoch 17/20
1407/1407 [==============================] - 101s 72ms/
    step - loss: 0.3822 - accuracy: 0.8803 - val_loss:
    0.5393 - val_accuracy: 0.8242
```

6.2.6.3 Accuracy and Loss Estimation with VGG-16

The accuracy and the loss of the implemented VGG-16 model are visualized using the matplot library. The function plot() is used with the training and validation for accuracy. The accuracy plot is shown in Figure 6.15. It can be seen that the training accuracy of the model is 88.03%, and validation accuracy is 82.42%. Hence, the validation is on par with the training. Similarly, the loss of the model is shown in Figure 6.16. It can be seen that the validation loss is less compared to the training loss. Therefore, the model has met the requirements for the generalization of the model.

```
plt.plot(training_history.history["accuracy"],label="train")
plt.plot(training_history.
history["val_accuracy"],label="val")
```

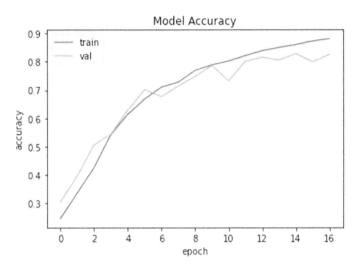

FIGURE 6.15
Accuracy of VGG-16.

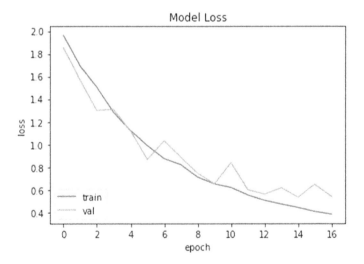

FIGURE 6.16
Loss of VGG-16.

```
plt.title("Model Accuracy")
plt.ylabel('accuracy')
plt.xlabel('epoch')
plt.legend(loc="upper left")
```

```
plt.plot(training_history.history["loss"],label="train")
plt.plot(training_history.history["val_loss"],label="val")
plt.title("Model Loss")
plt.ylabel('loss')
plt.xlabel('epoch')
plt.legend(loc="lower left")
```

6.3 YoloV3

Object classification has been used in AI systems to identify the objects that are of interest to a specific class. The identification and grouping of objects help to identify the different objects that are present in a scene. The applications of object detection span from crime scene analysis, traffic analysis and others. YoloV3 (You Only Look At Once Version 3) is one of the widely used networks for Object detection [4]. In this section, we explain the architecture details of YoloV3 and its implementation.

6.3.1 Architecture of YoloV3

YoloV3 uses basic convolutions as the baseline for object detection. It was designed by Joseph Redmon and Ali Farhadi using Keras and OpenCV libraries. The important characteristic of the YoloV3 network is the use of 1 × 1 convolutions. The name specification "You only look once" is because the prediction map has the same size as the feature map before it.

The architecture of YoloV3 is represented as shown in Figure 6.17. It consists of 53 convolution layers. Each layer has its own functionality and importance for the identification of objects. Each layer is followed by batch normalization and a leaky ReLU function. A stride of 2 is used to obtain the down-sampled feature maps with no pooling. For example, for an image of size 416 × 416, we obtain the feature map output of 13 × 13 with stride of 32. The implementation of the YoloV3 network is discussed in the next section.

6.3.2 Implementation

In this section, the implementation of the YoloV3 architecture is discussed with OpenCV and Keras. The dataset considered for the implementation is sample videos from YouTube. These videos are bundled as an input first and then the implementation is carried out.

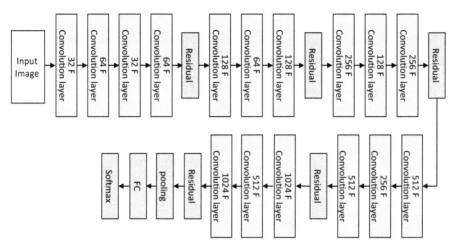

FIGURE 6.17
YoloV3 architecture.

6.3.3 Imports

The necessary libraries that are needed for the imports are numpy, opencv and others, as shown in the code snippet below.

```
import numpy as np
import cv2 as cv
import subprocess
import os
```

6.3.4 Reading the Dataset

The dataset that is considered for this example is a sample of YouTube videos. The script used for the usage of YouTube videos is shown below. The module pytube is used for the extraction of videos from YouTube. In this example, a list of links is used for the extraction of the videos. The link is passed to the streaming API for the download. In the next section, the object detection of the YoloV3 is discussed.

```
from pytube import YouTube

#the folder in which the videos are to be saves
SAVE_PATH = "./data/"

#links of the videos to be downloaded, seperated by ,
links=[ ]
```

```
for link in links:
    try:
    # object creation using YouTube
    # which was imported in the beginning
    yt = YouTube(link)
    except:
    #to handle exception
    print(f"Error couldn't download video: {link} ")
    try:
    # downloading the video
    stream = yt.streams.first()
    stream.download(SAVE_PATH)
    except:
    print("Error! couldnt download")

print('Finished operation!')
```

6.3.5 Object Detection using YoloV3

A class named 'Detector' is used for drawing the bounded boxes around the object identified using the Yolo network. The function draw_boxes() is used for this purpose with opencv function rectangle() based on the x, y positions of the objects. The function generate_boxes_confidences_classids() is used for representation of the text of the objects detected using the class ids and confidence intervals. The next section discusses the implementation of pre-trained Yolo network for object detection.

```
class Detector:

    def draw_boxes(self, img, boxes, confidences, classids,
        idxs, colors, labels):
        # if there is at least 1 detected objects
        if len(idxs) > 0:
            for i in idxs.flatten():
                # Get the coordinates for the boxes of the
                    detected object
                x, y = boxes[i][0], boxes[i][1]
                w, h = boxes[i][2], boxes[i][3]

                # Gives each class a distinct color
                    (randomly generated below)
                color = [int(c) for c in colors
                    [classids[i]]]

                # Draw a rectangle box  and assign the
                    label to the image
                cv.rectangle(img, (x, y), (x+w, y+h),
                    color, 2)
```

```python
                text = "{}: {:4f}".format(labels
                    [classids[i]], confidences[i])
                cv.putText(img, text, (x, y-5), 0, 0.5,
                    color, 2)

        return img


    def generate_boxes_confidences_classids(self, outs,
        height, width, tconf):
        boxes = []
        confidences = []
        classids = []

        for out in outs:
            for detection in out:

                # Get the scores, classid, and the
                    confidence of the prediction
                scores = detection[5:]
                classid = np.argmax(scores)
                confidence = scores[classid]

                # Consider only the predictions that are
                    above a certain confidence level
                if confidence > tconf:
                    # TODO Check detection
                    box = detection[0:4] * np.array([width,
                        height, width, height])
                    centerX, centerY, bwidth, bheight =
                        box.astype('int')

                    # Using the center x, y coordinates to
                        derive the top
                    # and the left corner of the bounding
                        box
                    x = int(centerX - (bwidth / 2))
                    y = int(centerY - (bheight / 2))

                    # Append the detected boxids,
                        confidence, classid to the
                        respective lists
                    boxes.append([x, y, int(bwidth),
                        int(bheight)])
                    confidences.append(float(confidence))
                    classids.append(classid)

        return boxes, confidences, classids
```

```
def detect_image(self ,net, layer_names, height, width,
    img, colors, labels, default_obj,
         boxes=None, confidences=None, classids=None,
idxs=None, infer=True):

    if infer:
        # Generate blob from the input image
        blob = cv.dnn.blobFromImage(img, 1 / 255.0,
            (416, 416), swapRB=True, crop=False)

        # Perform a forward pass of the blob on the
            yolo model
        net.setInput(blob)

        # Getting the outputs from the output layers
        outs = net.forward(layer_names)

        # Get the boxes, confidences, and classIDs of
            detected objects
        boxes, confidences, classids = self.generate_
            boxes_confidences_classids(outs, height,
            width, default_obj.confidence)

        # Apply Non-Max Suppression to suppress
            overlapping bounding boxes
        idxs = cv.dnn.NMSBoxes(boxes, confidences,
            default_obj.confidence, default_obj.
            threshold)

    # Draw boxes on the image
    img = self.draw_boxes(img, boxes, confidences,
        classids, idxs, colors, labels)

    return img, classids

detector_obj = Detector() # create a object of the
Detector class
```

A class named "SetDefault" is used for storing the pre-trained weights and models of YoloV3 network. The model is loaded using the function load_ net() with the module readNetFromDarknet(). The input and output paths are loaded in the class. The output folder is used for storing the images with the detected objects. It is followed by loading the video and running with a while loop to run object detection on each frame of the video and then saving the video after drawing boxes on detected boxes.

The different types of objects detected using the Yolo network are shown in the figures. The figure shows object detection using the Yolo network for electronics domain. It can be seen that objects detected are a book, monitor, cup and others. Similarly, Figures 6.18–6.22 show object detection using Yolo for household items. It can be seen here that one of the objects is represented as sofa and another as chair. This is because of the less confidence level and low light. The figure shows object detection using Yolo for automobile. Here also the label 'truck' is used though the visual might not be that of a truck. It

FIGURE 6.18
Object detection using YoloV3 for electronics.

FIGURE 6.19
Object detection using YoloV3 for household.

FIGURE 6.20
Object detection using YoloV3 for automobile.

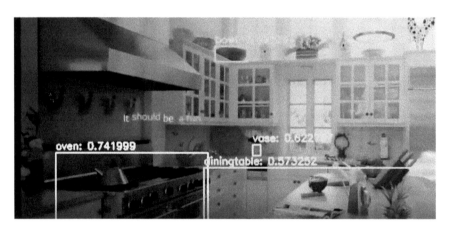

FIGURE 6.21
Object detection using YoloV3 for kitchen.

is because of the dimensions added to the weights of the Yolo network. The objects are identified similarly in the figures for automobile and sports.

```
class SetDefault:

  def __init__(self):

    self.model_path = "./yolov3-coco/"
    self.weights_path= "./yolov3-coco/yolov3.weights"
    self.config_path= "./yolov3-coco/yolov3.cfg"
    self.label_path = "./yolov3-coco/coco-labels"
```

FIGURE 6.22
Object detection using YoloV3 for sports.

```
    self.confidence= 0.5
    self.threshold =0.3
    self.file_name= "Basketball.mp4"
    self.input_folder= "./INPUT/"
    self.output_folder= "./OUTPUT/"

  def load_labels(self):
    labels= open(self.label_path).read().strip().
        split('\n')
    return labels

  def load_net(self):

    net= cv.dnn.readNetFromDarknet(self.config_path,
        self.weights_path)
    return net

  def get_layer_names(self, darknet ):

    layer_names = darknet.getLayerNames()
    layer_names = [layer_names[i[0] - 1] for i in
        darknet.getUnconnectedOutLayers()]
    return layer_names

default_obj = SetDefault()
labels= default_obj.load_labels()
net= default_obj.load_net()
colors = np.random.randint(0, 255, size=(len(labels), 3),
    dtype='uint8')
layer_names= default_obj.get_layer_names(net)
```

```
input_video_path= default_obj.input_folder + default_obj.
    file_name
output_video_path= default_obj.output_folder + default_
    obj.file_name

vid = cv.VideoCapture(input_video_path)
height, width = None, None
writer = None

while vid.isOpened():

    grabbed, frame = vid.read()
    if frame is None:
        break
    if not grabbed:
        break
    if width is None or height is None:
        height, width = frame.shape[:2]
    if writer is None:
        fourcc = cv.VideoWriter_fourcc(*"MJPG")
        writer = cv.VideoWriter(output_video_path,
            fourcc, 30, (frame.shape[1], frame.shape[0]),
            True)
    frame, classids = detector_obj.detect_image(net,
        layer_names, height, width, frame, colors, labels,
        default_obj)

    if frame is not None:
        print(classids)
        writer.write(frame)

print ("Finishing operations!!..")
writer.release()
vid.release()
```

References

1. Alom, M. Z., Taha, T. M., Yakopcic, C., Westberg, S., Sidike, P., Nasrin, M. S., ... & Asari, V. K. (2018). The history began from alexnet: A comprehensive survey on deep learning approaches. arXiv preprint arXiv:1803.01164.
2. https://www.cs.toronto.edu/~kriz/cifar.html
3. Simonyan, K., & Zisserman, A. (2014). Very deep convolutional networks. In *International Conference on Machine Learning and Applications (IEEE ICMLA'15)*, Anaheim, CA, USA.
4. Redmon, J., & Farhadi, A. (2018). Yolov3: An incremental improvement. arXiv preprint arXiv:1804.02767.

Exercises

1. Describe the functioning of the layers in AlexNet.
2. What is one-hot encoding? How many classes are there in CIFAR-10 dataset?
3. What is early stopping used for?
4. Compare and contrast the architecture of VGG-16 and Yolov3 with use cases.

7

Training Neural Networks on Cloud

7.1 Introduction

Neural networks form the basis of deep learning applications. We have seen various types of neural networks in the previous chapters with examples. However, the important part of the deep learning application is to prepare the data along with neural network and make it available on a cloud platform. Google Cloud and other components were introduced in the first sections of the book. In this chapter, the focus is on developing a simple convolutional neural network (CNN) with TensorFlow using Google Cloud Platform (GCP). It includes the sections of creating a virtual machine (VM) till the deployment.

7.2 Create a TensorFlow Application

Google Cloud Platform's user-managed notebooks provide us with notebooks along with a suite of deep learning suites preinstalled [1].

Step 1: Figure 7.1 shows different steps to create a notebook instance.

Step 2: Import the necessary function.

```
1. import numpy as np
2. import matplotlib.pyplot as plt
3. import tensorflow as tf
4. from random import randint
5. from sklearn.model_selection import train_test_split
```

Step 3: Load the CIFAR-10 dataset [2], consecutively splitting into its respective train and test data.

DOI: 10.1201/9781003215974-9

FIGURE 7.1
Create a notebook instance.

```
1. (train_images,train_lables),(test_images,test_
   lables) = tf.keras.datasets.cifar10.load_data()
2. no_of_train_images = train_images.shape[0]
3. image_shape = train_images.shape[1:]
```

```
4. print("No of Training images =",no_of_train_
   images,"each of shape =",image_shape)
5. print("No of Test images =",test_images.
   shape[0],"each of shape =",test_images.shape[1:])
```

Step 4: Create a class list of categories.

```
1. classes = list("airplane,automobile,bird,cat,deer,
   dog,frog,horse,ship,truck".split(","))
2. print(classes)
```

Step 5: Randomly display four images along with their labels and classes as shown in Figure 7.2

```
1. for i in range(4):
2.   k = randint(0,no_of_train_images)
3.   plt.subplot(2,2,i+1)
4.   plt.imshow(train_images[k])
5.   plt.title("label = "+str(train_lables[k]
          [0])+",class = "+classes[train_lables[k][0]])
6.   plt.tight_layout()
7.
```

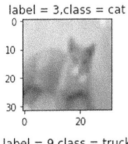
label = 3,class = cat

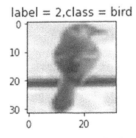
label = 2,class = bird

label = 9,class = truck

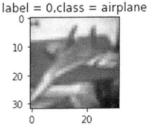
label = 0,class = airplane

FIGURE 7.2
Images with labels and classes.

Step 6: Convert labels from numerical to categorical for training purpose.

```
1. train_lables = tf.keras.utils.
   to_categorical(train_lables,num_classes=10)
2. test_lables = tf.keras.utils.
   to_categorical(test_lables,num_classes=10)
```

Step 7: Observe the training image format as shown in Figure 7.3

```
1. plt.imshow(train_images[1])
2. one_hot = train_lables[1]
3. lable = np.argmax(one_hot)
4. plt.title(f"One hot encoding = {one_hot}, label =
   {lable}, class = {classes[lable]}")
```

Step 8: Use the inbuilt Scikit function train_test_split() to split the train-
        ing images itself to train and validate data.

```
1. train_images, val_images, train_lables, val_lables
   = train_test_split(train_images, train_lables,
   test_size=0.1)
```

Step 9: Perform data augmentation using `ImageDataGenerator()`.

```
1. train_datagen = tf.keras.preprocessing.image.
   ImageDataGenerator(rescale=1./255, width_shift_
   range=0.1, height_shift_range=0.1, zoom_range=0.1,
   shear_range=0.1, rotation_range=2)
2. train_datagen.fit(train_images)
```

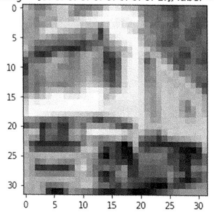

One hot encoding = [0. 0. 0. 0. 0. 0. 0. 0. 0. 1.], label = 9, class = truck

FIGURE 7.3
Trained image format.

```
3. training_set = train_datagen.flow(train_images,
   train_lables, batch_size=32)
4.
5. val_datagen = tf.keras.preprocessing.image.ImageDa
   taGenerator(rescale=1./255)
6. val_datagen.fit(val_images)
7. val_set = val_datagen.flow(val_images, val_lables,
   batch_size=32)
8.
```

Step 10: Build the Alexnet architecture.

Usually Alexnet starts with a Conv layer having a filter size of 11 × 11 for an input image of size 227 × 227. But since the size of CIFAR-10 images is 32 × 32, the filter dimensions are reduced at each layer. Even the number of dense units at the end is reduced. But the overall architecture is followed.

```
1. alexnet_model = tf.keras.models.Sequential(
2.    [
3.        tf.keras.layers.Conv2D(filters=32,kernel_
             size=3,activation="relu",input_shape=
             image_shape),
4.        tf.keras.layers.MaxPool2D
             (pool_size=(2,2),strides=1),
5.        tf.keras.layers.Conv2D(filters=32,kernel_size
             =3,padding="same",activation="relu"),
6.        tf.keras.layers.MaxPool2D
             (pool_size=(2,2),strides=1),
7.        tf.keras.layers.Conv2D(filters=64,kernel_size
             =3,padding="same",activation="relu"),
8.        tf.keras.layers.Conv2D(filters=64,kernel_size
             =3,padding="same",activation="relu"),
9.        tf.keras.layers.Conv2D(filters=64,kernel_size
             =3,padding="same",activation="relu"),
10.        tf.keras.layers.
             MaxPool2D(pool_size=(2,2),strides=1),
11.        tf.keras.layers.Flatten(),
12.        tf.keras.layers.Dense(units=512,activation="
             relu"),
13.        tf.keras.layers.Dropout(rate=0.2),
14.        tf.keras.layers.Dense(units=256,activatio
             n="relu",kernel_regularizer=tf.keras.
             regularizers.l2(0.001)),
15.        tf.keras.layers.Dropout(rate=0.2),
16.        tf.keras.layers.Dense(10,activation="softmax")
17.    ]
18.    )
```

1.　alexnet_model.summary()

Model: "sequential"

Layer (type)	Output Shape	Param #
conv2d (Conv2D)	(None, 30, 30, 32)	896
max_pooling2d (MaxPooling2D)	(None, 29, 29, 32)	0
conv2d_1 (Conv2D)	(None, 29, 29, 32)	9248
max_pooling2d_1 (MaxPooling2	(None, 28, 28, 32)	0
conv2d_2 (Conv2D)	(None, 28, 28, 64)	18496
conv2d_3 (Conv2D)	(None, 28, 28, 64)	36928
conv2d_4 (Conv2D)	(None, 28, 28, 64)	36928
max_pooling2d_2 (MaxPooling2	(None, 27, 27, 64)	0
flatten (Flatten)	(None, 46656)	0
dense (Dense)	(None, 512)	23888384
dropout (Dropout)	(None, 512)	0
dense_1 (Dense)	(None, 256)	131328
dropout_1 (Dropout)	(None, 256)	0
dense_2 (Dense)	(None, 10)	2570

Total params: 24,124,778
Trainable params: 24,124,778
Non-trainable params: 0

Step 11: Compile the model using Adam optimizer and calculate loss using categorical_crossentropy.

1.　alexnet_model.compile(optimizer=tf.keras.
optimizers.Adam(learning_rate=0.001), loss=
'categorical_crossentropy', metrics=['accuracy'])

Step 12: Train the model.

The Alexnet model is trained using the training set and validated against the validation set for 20 epochs with 32 images trained per batch. But the training will be stopped mid-way if there is no increase in validation accuracy for two consecutive epochs.

```
1. training_history = alexnet_model.fit
                        (training_set,
2.                       batch_size=32,
3.                       epochs=20,
4.                       validation_data = val_set,
5.                       callbacks = [tf.keras.
                            callbacks.EarlyStopping
                            (monitor='val_accuracy',
                            patience=2,restore_
                            best_weights=True)]
6.                      )
7.
```

Step 13: Plot the accuracy history while training the model as shown in Figure 7.4.

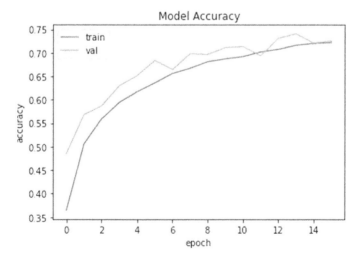

FIGURE 7.4
Model accuracy.

```
1. plt.plot(training_history.history["accuracy"],labe
   l="train")
2. plt.plot(training_history.history["val_accuracy"],
   label="val")
3. plt.title("Model Accuracy")
4. plt.ylabel('accuracy')
5. plt.xlabel('epoch')
6. plt.legend(loc="upper left")
```

Step 14: Plot the model loss while training the model as shown in Figure 7.5.

```
1. plt.plot(training_history.history["loss"],label=
   "train")
2. plt.plot(training_history.
   history["val_loss"],label="val")
3. plt.title("Model Loss")
4. plt.ylabel('loss')
5. plt.xlabel('epoch')
6. plt.legend(loc="lower left")
```

Step 15: Evaluate the model.

```
1. test_images_1 = test_images/255
2. alexnet_model.evaluate(x=test_images_1,
   y=test_lables)
```

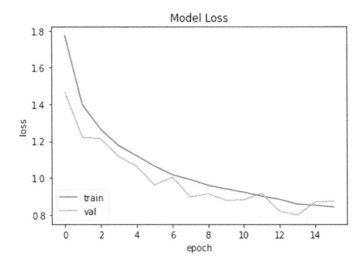

FIGURE 7.5
Model loss.

Step 16: Save the model in your preferred bucket.

```
1. MODEL = 'cifar-10'
2. MODEL_SAVE_BUCKET = 'gs://cifar-10-model'
3. alexnet_model.save(MODEL_SAVE_BUCKET +
   'saved_model')
```

Observe the bucket location after execution, which is shown in Figure 7.6.

Now that we have a trained model stored in our bucket, we can then create a model in the artificial intelligence (AI) platform and link the model as its first version. Figure 7.7 illustrates the process of creating a model in the AI platform and linking the model.

FIGURE 7.6
Model saving in the bucket.

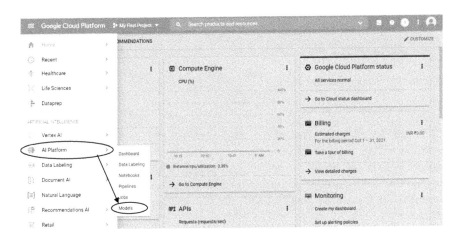

FIGURE 7.7
Creation of model in the AI platform and linking the model.

(*Continued*)

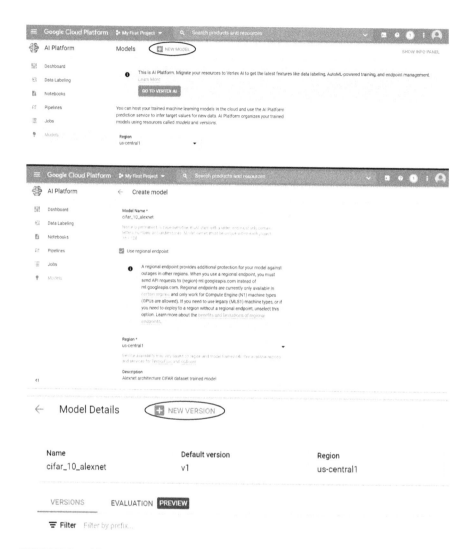

FIGURE 7.7 (CONTINUED)

Fill the details as shown above and provide the bucket URI where the model is saved. Ensure that the online prediction deployment will be scaled manually (*Manual Scaling*); the rest can be filled with your preferred choice of machine and number of nodes. Our main focus is the batch prediction deployment this time; you can still try out the online prediction with the same input.

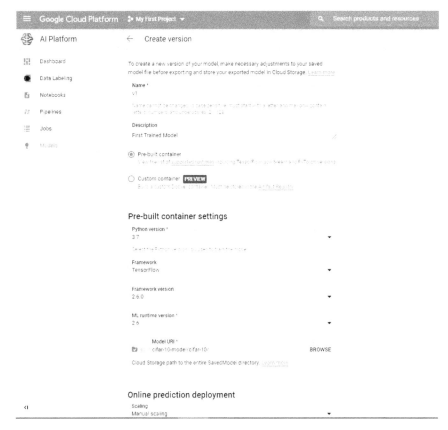

FIGURE 7.7 (CONTINUED)

7.3 Running using Single-Worker Instance

To run a batch prediction with a single worker, we would need a test dataset. The test dataset used to benchmark the difference between single-worker and multi-worker prediction comprises ten high-quality images based on the ten classes of classification.

To run a model by modifying the number of workers is to make the model in AI platform and then import the model to Vertex AI, which is shown in Figure 7.8. After importing the model, we can run the batch predictions.

We have to type the same model name created in the AI platform and also make sure that the region is also the same as the earlier created model shown in Figure 7.9.

The same parameters have to be given while importing the model.

FIGURE 7.8
Importing Vertex AI.

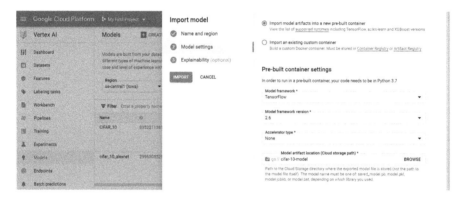

FIGURE 7.9
Configuring created AI model.

After specifying the parameters, we can import the model to perform batch prediction.

The most important parameter for batch prediction is to provide the same type of input given during training while running the batch prediction. The input for batch prediction must be stored in a bucket with the same region as the model, as shown in Figure 7.10.

To test the batch prediction, we must initially set the correct model name. The source path is the file which input file. The input file can be of type CSV, JSONL or TFRecord. Then, we must select the correct output location, as shown in Figure 7.11.

The most important parameter for single-worker instance is to specify that the number of compute nodes is 1.

After specifying the parameters, we run the prediction and we get the output shown in Figure 7.12.

We have provided 1 image to test the batch prediction and the input is in the form of Numpy array(JSON). The output is stored in a bucket in the form of JSON file.

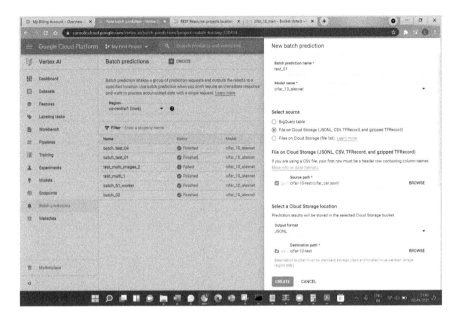

FIGURE 7.10
Creation of new batch prediction.

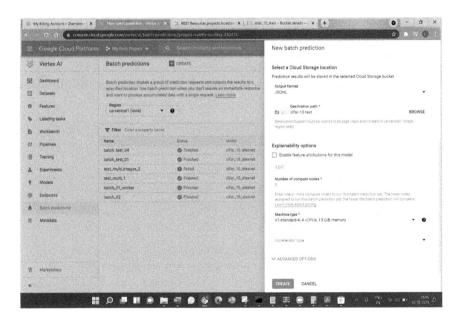

FIGURE 7.11
Configuring created batch prediction.

Vertex AI		← batch_test_01	
Dashboard		ID	3288144498445516800
Datasets		Model	cifar_10_alexnet
		Objective	Custom
Features		Import location	gs://cifar_10_train/cifar_6.jsonl
Labeling tasks		Total items	1
		Predicted items	1
Workbench		Created	Nov 02, 2021 at 08:37PM
		Updated	Nov 02, 2021 at 08:56PM
Pipelines		Elapsed time	16 min 26 sec
Training		Status	Finished
		Export location	gs://cifar-10-test/prediction-cifar_10_alexnet-2021_11_02T08_07_39_775Z
Experiments			
Models			
Endpoints			
Batch predictions			
Metadata			

FIGURE 7.12
Worker instance.

7.4 Running using Distributed Prediction

While creating the version in AI platform, the most important setting is to allow manual scaling.

Manual scaling allows user to change the number of workers while giving the predictions. Figure 7.13 shows online prediction deployment scenario.

We test another image using four workers with the same machine type we used in the single-worker instance.

N1 Standard-4, which is shown in Figure 7.14. We have set the number of workers to 4.

Online prediction deployment

Keeping a minimum number of nodes running all the time will avoid dropping requests due to nodes initialization after the service has scaled down. This setting can increase cost, as you pay for the nodes even when no predictions are served.

FIGURE 7.13
Online prediction deployment.   *(Continued)*

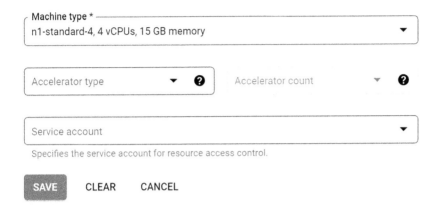

FIGURE 7.13 (CONTINUED)

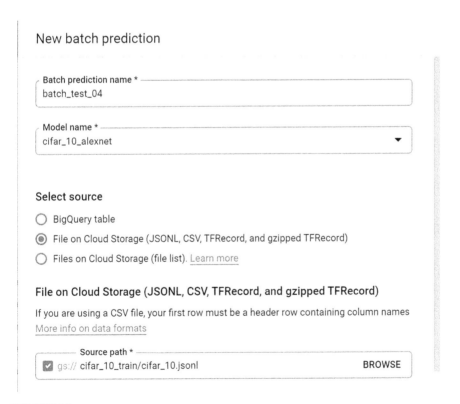

FIGURE 7.14
Four-worker nodes. *(Continued)*

Select a Cloud Storage location

Prediction results will be stored in the selected Cloud Storage bucket

Output format

JSONL   ▼

Destination path *

☑ gs:// cifar-10-test   BROWSE

Destination bucket must be standard storage class and located in us-central1 (single region only)

Explainability options

☐ Enable feature attributions for this model

EDIT

Number of compute nodes *

4

Enter one or more compute nodes to run this batch prediction job. The more nodes assigned to run this batch prediction job, the faster the batch prediction will complete. Learn more about pricing

Machine type *

n1-standard-4, 4 vCPUs, 15 GiB memory   ▼   ❓

Accelerator type   ▼

∨ ADVANCED OPTIONS

FIGURE 7.14   (CONTINUED)

After running the batch prediction with four workers, we get the following output, which is shown in Figure 7.15.

FIGURE 7.15
Output of four workers.

7.5 Hyperparameters and Optimization

For testing the hyperparameters, we use a higher-memory CPU N1-highmem-8. We are testing the same images that we used for the single and distributed worker instances. Figure 7.16 shows hyperparameters and optimization.

After running this batch prediction, we get the output, which is shown in Figure 7.17.

Now we test the predictions with four workers using the higher-memory machine type. Figure 7.18 shows the four workers using higher-memory machine type.

Figure 7.19 shows the output of the execution of four workers using higher-memory machine types.

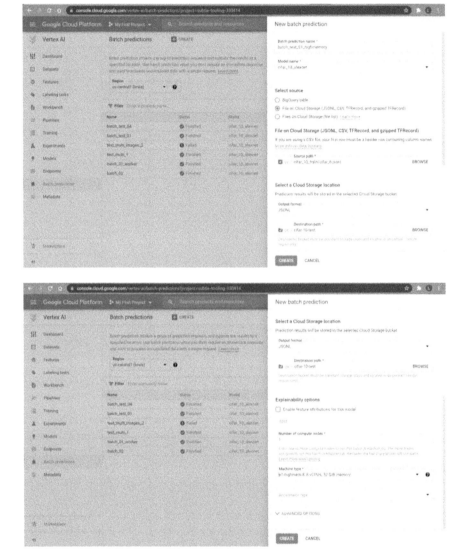

FIGURE 7.16
Hyperparameters and optimization.

← batch_test_01_highmemory

ID	4152378230063497216
Model	cifar_10_alexnet
Objective	Custom
Import location	gs://cifar_10_train/cifar_6.jsonl
Total items	1
Predicted items	1
Created	Nov 03, 2021 at 12:13PM
Updated	Nov 03, 2021 at 12:31PM
Elapsed time	15 min 23 sec
Status	Finished
Export location	gs://cifar-10-test/prediction-cifar_10_alexnet-2021_11_02T23_43_06_792Z

FIGURE 7.17
Output of batch prediction in hyperparameters and optimization.

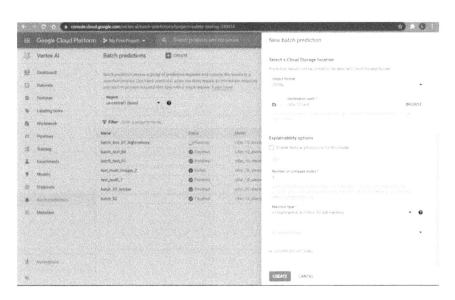

FIGURE 7.18
Four workers using higher-memory machine type. *(Continued)*

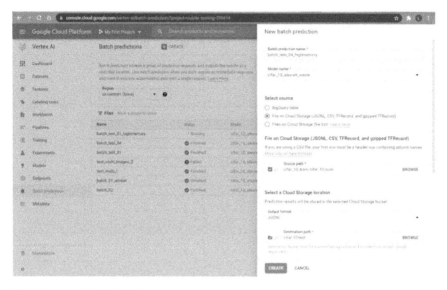

FIGURE 7.18 (CONTINUED)

← batch_test_04_highmemory

ID	2839930782405885952
Model	cifar_10_alexnet
Objective	Custom
Import location	gs://cifar_10_train/cifar_10.jsonl
Total items	1
Predicted items	1
Created	Nov 07, 2021 at 11:06AM
Updated	Nov 07, 2021 at 11:25AM
Elapsed time	15 min 31 sec
Status	Finished
Export location	gs://cifar-10-test/prediction-cifar_10_alexnet-2021_11_06T22_36_24_087Z

FIGURE 7.19
Output of execution of four workers using higher-memory machine types.

FIGURE 7.20
Prediction output – GCP.

7.6 Deployment and Prediction

While giving the batch prediction, we specify the output bucket to which the predictions are saved.

Figure 7.20 shows the sample prediction output given by GCP. We see a 99.3489% prediction that the above image belongs to the first class, which is airplane.

References

1. https://cloud.google.com/deep-learning-vm/docs/tensorflow_start_instance
2. https://www.cs.toronto.edu/~kriz/cifar.html

Exercises

1. Where to enable APIs? What is the use of Compute Engine API?
2. What are some parameters to be specified while creating virtual machines on Google Cloud Platform?
3. What is the use of DockerFile for creating a model on GCP?
4. How does manual scaling differ from automatic scaling in online prediction
5. What is the use of Google Container Registry API?

Part III

Cloud-based Applications of Multi-Modal Analytics

8

Image Analytics

DOI: 10.1201/9781003215974-11

8.1 Introduction

The era of smartphones has enabled the explosion of image data through a camera and other photo applications. There are various facial recognition applications that are used in the smartphone for various purposes [1]. Image data can be used for other applications like object detection, crime detection and others. The different types of neural networks like CNN, RNN and others can be used for image analytics and classification. In this chapter, an example shows how to create a basic image classification model with TensorFlow's built-in "Rock-Paper-Scissor" picture dataset and deploy it on Google Cloud Platform.

8.2 Data Preparation with Google Cloud

'Rock-Paper-Scissor' dataset contains different images of the hands showing a 'Rock' or 'Paper' or 'Scissor' [2]. The data is imported using tensorflow_dataset library. Data sharding is done with a fixed batch size of 64.

8.3 Image Recognition with Google Cloud and TensorFlow

8.3.1 Create a TensorFlow Application

The docker file containing the required scripts to import the dataset and train the model must be built on an environment with the appropriate versions of library supporting it.

For this, we use the Compute Engine service to create virtual machine (VM) instance. Our initial step is to enable the Compute Engine API. Figure 8.1 shows the process of enabling Compute Engine API.

Once the Compute Engine API is enabled, we must now create a VM instance. Figure 8.2 shows the VM instance creation.

Enter the instance name and the machine configuration as shown in the image below. Figure 8.3 shows the same.

Now change the Booting OS to Ubuntu 18.04LTS with the default boot disk type and a sufficient space of 20Gb. Figures 8.4 and 8.5 show boot disk configuration of the VM instance.

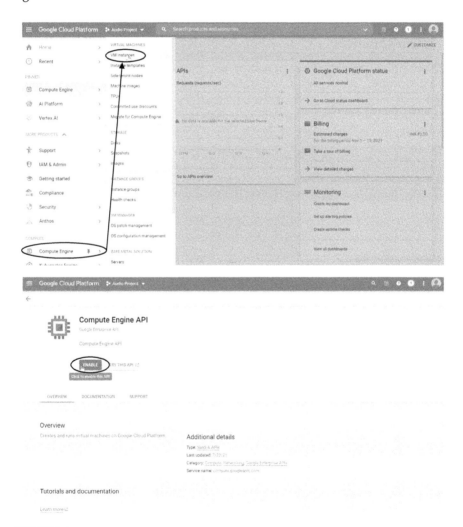

FIGURE 8.1
Compute Engine API enabling.

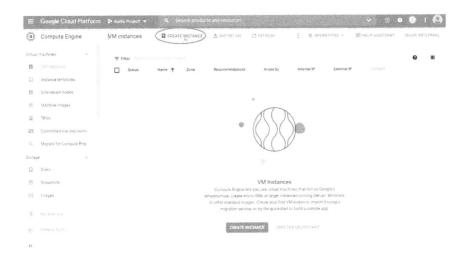

FIGURE 8.2
Creation of VM instance.

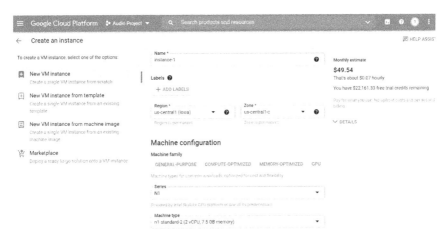

FIGURE 8.3
VM instance – machine configuration.

Ensure that 'Allow full access to all cloud APIs' is selected so that the VM instance can interact with the cloud storage as well as the Cloud Container Registry API.

A docker file container once built must be stored. Google Cloud provides a container registry to manage docker containers. Let's enable the Container Registry API. Figure 8.6 shows the procedure for Google Container Registry API.

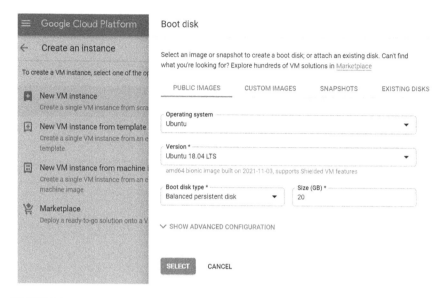

FIGURE 8.4
VM instance – Boot disk configuration -1.

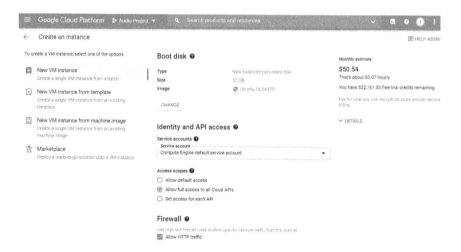

FIGURE 8.5
VM instance – Boot disk configuration -2.

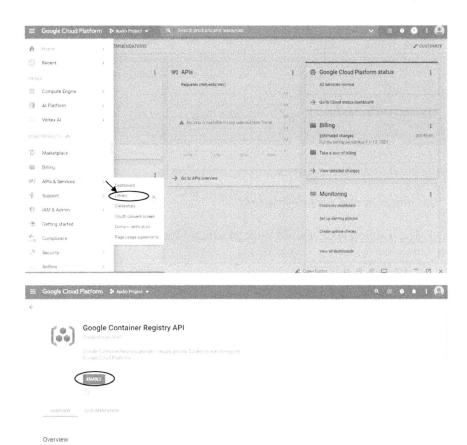

FIGURE 8.6
Google Container Registry API.

Once the Container Registry API is enabled. Our next step is to SSH into the VM instance created and build the docker file.

After connecting to VM, verify whether the OS is up-to-date.

```
kesevanbalaji@instance-1:~$ sudo apt update
Hit:1 http://us-centrall.gce.archive.ubuntu.com/ubuntu bionic InRelease
Hit:2 http://us-centrall.gce.archive.ubuntu.com/ubuntu bionic-updates InRelease
Hit:3 http://us-centrall.gce.archive.ubuntu.com/ubuntu bionic-backports InRelease
Hit:4 http://security.ubuntu.com/ubuntu bionic-security InRelease
Reading package lists... Done
Building dependency tree
Reading state information... Done
5 packages can be upgraded. Run 'apt list --upgradable' to see them.
```

Ensure that docker is installed using the following command as shown in Figure 8.7.

Training code and project structure

1. The training code should be organized in the following way cloud_ training folder:

    • trainer folder:
        • __init__.py
        • task.py
        • model.py
    • Dockerfile
    • config.yaml

2. task.py contains all the instructions to obtain the dataset, process it, create the model and run the training loop. It also specifies the distribution strategy to be used; in this case, it is the Multi Worker Mirrored Strategy.

    ```
    1. import tensorflow as tf
    2. import tensorflow_datasets as tfds
    3. import argparse
    4. import json
    5. import os
    6.
    ```

```
kesevanbalaji@instance-1:~$ sudo apt install docker.io
Reading package lists... Done
Building dependency tree
Reading state information... Done
docker.io is already the newest version (20.10.7-0ubuntu5~18.04.3).
The following package was automatically installed and is no longer required:
  libnumal
Use 'sudo apt autoremove' to remove it.
0 upgraded, 0 newly installed, 0 to remove and 2 not upgraded.
```

FIGURE 8.7
SSH into the VM instance to build docker file.

```
 7. from trainer.model import create_model
 8.
 9. REPLICA_BATCH_SIZE = 64
10.
11. def get_args():
12.
13.   parser = argparse.ArgumentParser()
14.   parser.add_argument(
15.     '--epochs',
16.     required=True,
17.     type=int)
18.   parser.add_argument(
19.     '--job-dir',
20.     required=True,
21.     type=str)
22.   args = parser.parse_args()
23.   return args
24.
25.
26. def preprocess_data(image, label):
27.
28.   image = tf.image.resize(image, (300,300))
29.   return tf.cast(image, tf.float32) / 255., label
30.
31.
32. def create_dataset(batch_size):
33.
34.   data, info = tfds.load(name='rock_
         paper_scissors', as_supervised=True,
         with_info=True)
35.   num_classes = info.features['label'].num_classes
36.   train = data['train'].map(preprocess_data,
37.                     num_parallel_calls=tf.data.
                        experimental.AUTOTUNE)
38.   train  = train.shuffle(1000)
39.   train  = train.batch(batch_size)
40.   train  = train.prefetch(tf.data.experimental.
         AUTOTUNE)
41.   options = tf.data.Options()
42.   options.experimental_distribute.auto_
         shard_policy = tf.data.experimental.
         AutoShardPolicy.DATA
43.   train_data = train.with_options(options)
44.   return train_data, num_classes
45.
46.
47. def _is_chief(task_type, task_id):
48.
49.   return task_type == 'chief'
```

```
50.
51.
52. def _get_temp_dir(dirpath, task_id):
53.
54.     base_dirpath = 'workertemp_' + str(task_id)
55.     temp_dir = os.path.join(dirpath, base_dirpath)
56.     tf.io.gfile.makedirs(temp_dir)
57.     return temp_dir
58.
59.
60. def write_filepath(filepath, task_type, task_id):
61.
62.     dirpath = os.path.dirname(filepath)
63.     base = os.path.basename(filepath)
64.     if not _is_chief(task_type, task_id):
65.        dirpath = _get_temp_dir(dirpath, task_id)
66.     return os.path.join(dirpath, base)
67.
68.
69. def main():
70.     args = get_args()
71.     strategy = tf.distribute.
            MultiWorkerMirroredStrategy()
72.     global_batch_size = REPLICA_BATCH_SIZE *
        strategy.num_replicas_in_sync
73.     train_data, number_of_classes =
        create_dataset(global_batch_size)
74.
75.     with strategy.scope():
76.        model = create_model(number_of_classes)
77.
78. model.fit(train_data, epochs=args.epochs)
79.
80. task_type, task_id = (strategy.cluster_resolver.
    task_type,
81.         strategy.cluster_resolver.task_id)
82.
83. write_model_path = write_filepath(args.job_dir,
    task_type, task_id)
84. model.save(write_model_path)
85.
86. if __name__ == "__main__":
87.     main()
```

3. model.py contains the model architecture that we'll be using. The model can also be specified in task.py, thus eliminating the need for model.py.

```
1.  import tensorflow as tf
2.
3.  def create_model(number_of_classes):
4.      base_model = tf.keras.applications.
            ResNet50(weights='imagenet', include_top=False)
5.      x = base_model.output
6.      x = tf.keras.layers.GlobalAveragePooling2D()(x)
7.      x = tf.keras.layers.Dense(1016,
            activation='relu')(x)
8.      predictions = tf.keras.layers.Dense(number_of_
            classes, activation='softmax')(x)
9.      model = tf.keras.Model(inputs=base_model.input,
            outputs=predictions)
10.
11.     model.compile(
12.         loss='sparse_categorical_crossentropy',
13.         optimizer=tf.keras.optimizers.Adam(0.0001),
14.         metrics=['accuracy'])
15.
16.     return model
17.
```

4. The write_filepath method inside the sample task.py will save the model to the cloud storage bucket automatically once the model has finished training, and is therefore essential to include in the program.

5. A dockerfile is a text document that contains all the commands a user could call on the command line to assemble an image. This is vital when it comes to building the image.

```
1.  # Specifies base image and tag
2.  FROM gcr.io/deeplearning-platform-release/
        tf2-gpu.2-5
3.  WORKDIR /root
4.
5.  # Copies the trainer code to the docker image.
6.  COPY trainer/ /root/trainer/
7.
8.  # Sets up the entry point to invoke the trainer.
9.  ENTRYPOINT ["python", "-m", "trainer.task"]
10.
```

6. config.yaml and its functioning is described in the configuring clusters section trainingInput:

```
1.  trainingInput:
2.  scaleTier: CUSTOM
```

```
 3. masterType: n1-standard-16
 4. masterConfig:
 5.    imageUri: gcr.io/arctic-ocean-331513/
          multicore:v1
 6. useChiefInTfConfig: true
 7. workerType: n1-standard-16
 8. workerCount: 1
 9. workerConfig:
10.    imageUri: gcr.io/arctic-ocean-331513/
          multicore:v1
```

Export the required variables to be used for building the docker file. Figure 8.8 shows different steps of building docker file.

1. **export   PROJECT_ID=$(gcloud   config   list   project   --format "value(core.project)")**

2. **export IMAGE_URI=gcr.io/$PROJECT_ID/{repo_name}:{tag}**

   Repo name will be the name of your object in the container registry, and tag will differentiate it from other similarly named containers. Example **export IMAGE_URI=gcr.io/$PROJECT_ID/multicore:v1**

```
kesevanbalaji@instance-1:~$ export PROJECT_ID=$(gcloud config list project --format "value(cor
e.project)")
kesevanbalaji@instance-1:~$ echo $PROJECT_ID
arctic-ocean-331513
kesevanbalaji@instance-1:~$ export IMAGE_URI=gcr.io/$PROJECT_ID/multicore:v1
kesevanbalaji@instance-1:~$ echo $IMAGE_URI
gcr.io/arctic-ocean-331513/multicore:v1
```

Now run the following command to

```
kesevanbalaji@instance-1:~$ sudo docker build -f Dockerfile -t $IMAGE_URI .
Sending build context to Docker daemon  136.2kB
Step 1/4 : FROM gcr.io/deeplearning-platform-release/tf2-gpu.2-5
---> 307b41b1aec7
Step 2/4 : WORKDIR /root
---> Using cache
---> 3d341c9faa07
Step 3/4 : COPY trainer/ /root/trainer/
---> Using cache
---> 81fbefb78441
Step 4/4 : ENTRYPOINT ["python", "-m", "trainer.task"]
---> Using cache
---> a4cc00b6b9d8
Successfully built a4cc00b6b9d8
Successfully tagged gcr.io/arctic-ocean-331513/multicore:v1
```

This builds the container, which is then pushed into the container registry through the following commands.

```
kesevanbalaji@instance-1:~$ gcloud auth configure-docker
WARNING: Your config file at [/home/kesevanbalaji/.docker/config.json] contains these credenti
al helper entries:

{
  "credHelpers": {
    "gcr.io": "gcloud",
    "us.gcr.io": "gcloud",
    "eu.gcr.io": "gcloud",
    "asia.gcr.io": "gcloud",
    "staging-k8s.gcr.io": "gcloud",
    "marketplace.gcr.io": "gcloud"
  }
}
Adding credentials for all GCR repositories.
WARNING: A long list of credential helpers may cause delays running 'docker build'. We recomme
nd passing the registry name to configure only the registry you are using.
gcloud credential helpers already registered correctly.
```

```
kesevanbalaji@instance-1:~$ sudo docker push $IMAGE_URI
The push refers to repository [gcr.io/arctic-ocean-331513/multicore]
5e223d8bb836: Layer already exists
5bb1aa5df10d: Layer already exists
f028010939aa: Layer already exists
dc99c4ea3a81: Layer already exists
37b508c5711b: Layer already exists
756ab564e194: Layer already exists
2ae86808a3d1: Layer already exists
1dccbdf9b557: Layer already exists
cfcbdbc2b748: Layer already exists
937ab8f29c2e: Layer already exists
5d417b2f7486: Layer already exists
d6a297a3e6e4: Layer already exists
6474a5e8117f: Layer already exists
fe498124ed57: Layer already exists
d5454704bb3d: Layer already exists
fb896ef24b4b: Layer already exists
5087113f67c8: Layer already exists
2a92857a1d48: Layer already exists
0ded97864c52: Layer already exists
b50bbaac3e32: Layer already exists
262ea1af4c10: Layer already exists
b420a468ca49: Layer already exists
608c205798d1: Layer already exists
0760cd6d4269: Layer already exists
fb4755c89c2a: Layer already exists
22cfb9034da6: Layer already exists
8bec4fbfce85: Layer already exists
3b129ca3db46: Layer already exists
64cb1a1930ab: Layer already exists
600ef5a43f1f: Layer already exists
8f8f0266f834: Layer already exists
v1: digest: sha256:c61734e62d67eb199d9dc04e7f8f63116e1fc26e57fdf1d5b95712d14def87c2 size: 6836
```

FIGURE 8.8
Process of building the docker file.

After the docker image is finished pushing, navigate to the container registry section and find the image listed, which is shown in Figure 8.9.

Now that we have the container stored in the registry, our next goal must be to train the model.

8.3.2 Running using Single-Worker/Distributed Instance

Training can be done using Google Cloud's AI Platform services. A training job can be submitted as a command or through GCP UI manually.

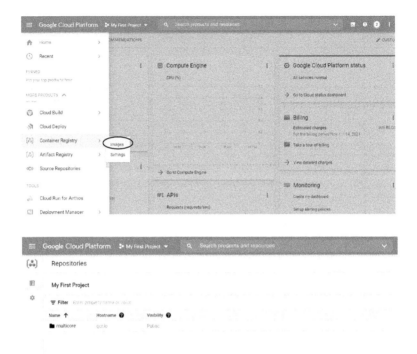

FIGURE 8.9
Container Registry.

Training on single worker and multiple workers:

1. The config.yaml file specifies which type of machine will act as the master and workers. It also sets the number of workers that will be used for the training job.

```
config.yaml  ×
  config.yaml  >  ...
  1    trainingInput:
  2      scaleTier: CUSTOM
  3      masterType: n1-standard-16
  4      masterConfig:
  5        imageUri: gcr.io/arctic-ocean-331513/multicore:v1
  6      useChiefInTfConfig: true
  7      workerType: n1-standard-16
  8      workerCount: 1
  9      workerConfig:
 10        imageUri: gcr.io/arctic-ocean-331513/multicore:v1
 11
```

2. In config.yaml, one machine is specified as a master, which will coordinate between the multiple worker machines.

3. The worker machines can be set to any machine type available on GCP in the region that will be running the training job. The number of workers can be adjusted according to the size of the training job.

4. The ImageUri specifies which docker image from the container registry can be used to complete the training job.

5. If training on a single worker, then the **workerCount:1**; if training on multiple workers the **workerCount** can be specified to any number.

1. To launch a training job on GCP, run the following command from the local project directory.

```
1. gcloud ai-platform jobs submit training {job_name} \
2.    --region europe-west2 \
3.    --config config.yaml \
4.    --job-dir gs://{gcs_bucket/model_dir} -- \
5.    --epochs 5 \
```

2. Replace {job_name} with a name for the training job.

3. The --region argument specifies the GCP region the training job would run in.

4. The --job-dir argument specifies which bucket and directory to store the final model in; the gcs_bucker name should be the same as the one initially created in the cloud storage section. The model_dir can be specified as any unique name; this directory will be automatically created in the specified GCP bucket.

5. The --epochs flag specifies the number of epochs for which training will be performed.

```
kesevanbalaji@cloudshell:~ (arctic-ocean-331513)$ gcloud ai-platform jobs submit training docker_train    --region europe-west2 --
config config.yaml    --job-dir gs://multi-worker --    --epochs 5 \
>
Job [docker_train] submitted successfully.
Your job is still active. You may view the status of your job with the command

  $ gcloud ai-platform jobs describe docker_train

or continue streaming the logs with the command

  $ gcloud ai-platform jobs stream-logs docker_train
jobId: docker_train
state: QUEUED
```

6. The training job takes about ten minutes to be queued. The status of the training job and logs can be viewed in the AI Platform/jobs section on the GCP console.

8.3.3 Hyperparameters and Optimization

To perform hyperparameter tuning or optimization, we can perform dropout regularization by changing the model parameters in model.py with a dropout rate of your choice.

8.3.3.1 Deployment and Prediction

Now that the model is trained and the resulting weights file (saved_model. pb) is saved to the given bucket, the next step is to import the model to Vertex AI platform to deploy and perform predictions.

Note: The predictions are made using batch prediction technique only; hence, it is not necessary to deploy the model. But to perform prediction using online prediction method, it is required to deploy the model to endpoint.

First, ensure that the Vertex AI platform is enabled.

Once enabled, proceed to the model section and import model from bucket, as shown in Figure 8.10.

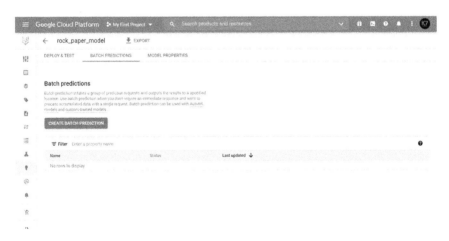

FIGURE 8.10
Model section and import model from bucket.

Enter the name of the model to be imported and set the region.

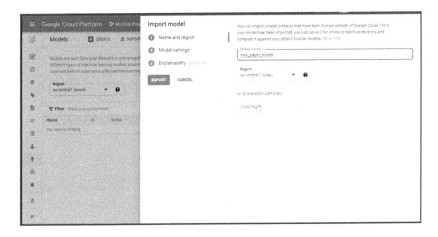

Next, select the option to *import model artifacts into a new prebuilt container*. Enter the details of versions of the library used in the training of this model and upload the cloud storage path.

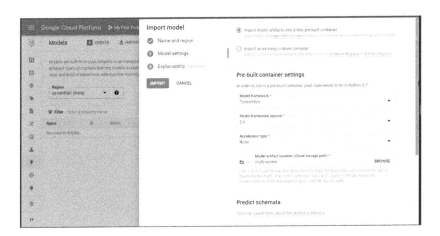

Once the model is imported, deploy the model to endpoint if online predictions need to be made.

Let us now perform batch predictions accordingly.

To do batch/online prediction, we must verify that the data used for prediction is the same as the data used to train the model. Because the model is trained with images in the form of a numpy array in this example, the prediction images must be transformed to a numpy array before being put in a JSON file.

Go to Cloud shell and create a python file with the following script.

```
1. import tensorflow as tf
2. import tensorflow_datasets as tfds
3. import numpy as np
4. import json
5.
6. DATASET_NAME = 'rock_paper_scissors'
7. name="scissor.jsonl"
8.
9. (dataset_train_raw, dataset_test_raw), dataset_
       info = tfds.load(
10.     name=DATASET_NAME,
11.     data_dir='tmp',
12.     with_info=True,
13.     as_supervised=True,
14.     split=[tfds.Split.TRAIN, tfds.Split.TEST],
15. )
16.
17. def preview_dataset(dataset):
18.     for features in dataset.take(1):
19.         (image, label) = features
20.         label = get_label_name(label.numpy())
```

```
21.              image=tf.image.resize(image,size=(150,
                    150))
22.              imm = image.numpy()
23.              lists = imm.tolist()
24.              json_str = json.dumps(lists)
25.              with open(name,'w') as test_file:
26.                  test_file.write(json_str)
27. preview_dataset(dataset_test_raw)
28. print("Saved Input Image in:"+name)
```

This script when executed will provide a JSONL file at your shell environment. Download the JSONL file and upload it to cloud storage bucket. This JSONL file will be used as the input for the batch prediction.

Create a batch prediction task, name it, and the node's machine type, including the number of nodes to do the prediction.

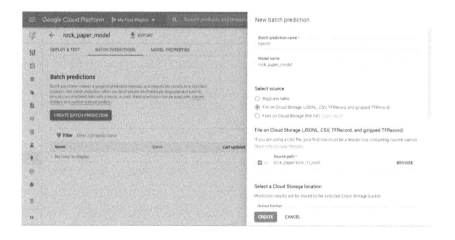

For the first batch prediction, use a single compute node with a machine type of your choice.

Note: To give a performance difference between a single node and several nodes, the machine type must be the same. Throughout the predictions, the *n1-standard-4,4 vCPUs,15 GiB* Memory machine type is utilized.

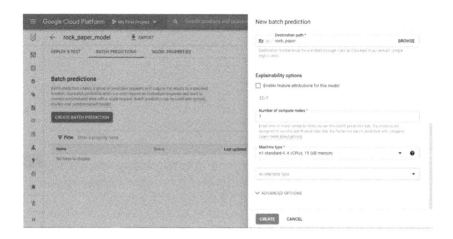

Create the batch prediction job. Now create a batch prediction with the same machine type but with four or more compute nodes.

The output of both predictions gets stored in the given cloud storage bucket. The prediction will be seen after the numpy array (Figure 8.11).

0.9803921580314636, 0.9803921580314636], [0.9794117212295532, 0.9794117212295532, 0.9794117212295532], [0.981372594833374, 0.981372594833374, 0.981372594833374],
[0.9784313440322876, 0.9784313440322876, 0.9784313440322876], [0.9794117212295532, 0.9794117212295532, 0.9794117212295532], [0.9823529720306396, 0.9823529720306396, 0.9823529720306396,
0.9823529720306396], [0.9823529124259949, 0.9823529124259949, 0.9823529124259949]]], "prediction": [0.868913472, 0.071821332, 0.0592652075]}

FIGURE 8.11
Creation of a batch prediction task.

TABLE 8.1

Results of Created Batch Prediction Task

Instance	Elapsed Time
Single-worker Instance	24 min 37 sec
Distributed Instance (4 instances)	20 min 53 sec

After executing both batch predictions with the same input, we can observe in Table 8.1.

References

1. Lin, M. S., Liang, Y., Xue, J. X., Pan, B., & Schroeder, A. (2021). Destination image through social media analytics and survey method. *International Journal of Contemporary Hospitality Management*. 33(6), 2219–2238.
2. https://www.tensorflow.org/datasets/catalog/rock_paper_scissors

Exercises

1. What is the general format of IMAGE_URI? What is the use of IMAGE_URI?
2. What is the function of a YAML file?
3. What are the two ways to take prediction after model is ready?
4. What are some parameters to be given as input during batch prediction?
5. What format is the prediction obtained? Where do you get the predictions?

9

Text Analytics

9.1 Introduction

Text analytics involves the process of converting unstructured text to structured and meaningful data using various techniques [1]. The different types of applications as a part of text analytics include sentiment analysis, category classification, spam classification and others [2]. Deep learning methods including RNN-based networks like LSTM and GRU are most useful for text analytics and its applications. In this chapter, we use neural networks to explore text analytics and how it can be deployed in GCP.

9.2 Data Preparation with Google Cloud

Text analytics is used for deeper insights, like identifying a pattern or trend from unstructured text. GloVe (Global Vectors) is a model for distributed word representation. This is achieved by mapping words into a meaningful space where the distance between words is related to semantic similarity. One benefit of GloVe is that it is the result of directly modeling relationships, instead of getting them as a side effect of training a language model. For example, in this chapter we will build a text analytics model to understand the mood of the user based on the words used. The dataset used is Glove 6B 300d [3] consisting of words with values for each kind of mood.

DOI: 10.1201/9781003215974-12

9.3 Speech Classification with Google Cloud and TensorFlow

9.3.1 Create a TensorFlow Application

We use the TensorFlow built-in dataset to run this model.

As this model needs a lot of memory, it is needed to create a VM (virtual machine) instance, rather than using the cloud shell, which would run out of disk space.

For creating the VM instance, user needs to enable the Compute Engine API.

On the navigation bar, navigate to APIs & Services and enable the Compute Engine API. Figure 9.1 shows the process of enabling Compute Engine API.

After enabling the API, navigate to Compute Engine/VM Instances. Figure 9.2 shows steps to configure GCP instance.

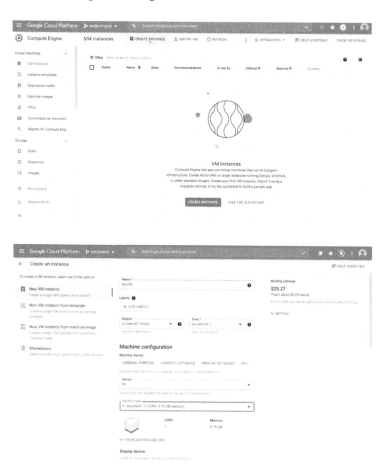

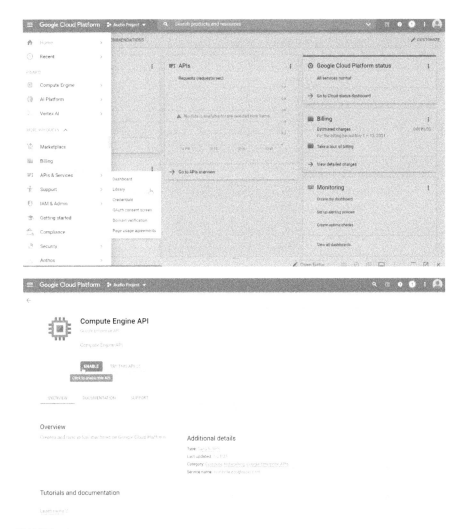

FIGURE 9.1
Steps to enable the Compute Engine API.

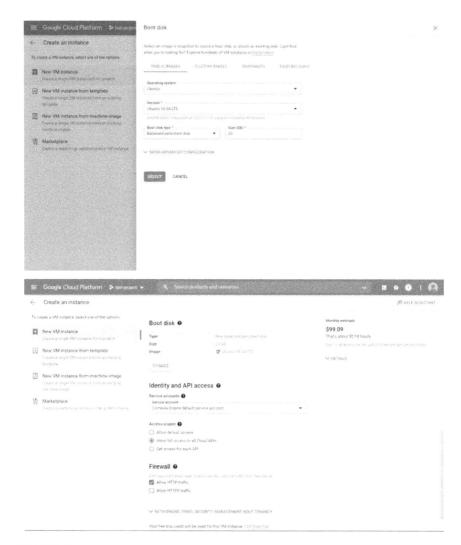

FIGURE 9.2
Steps to configure GCP instance.

User need to specify the instance name ending with a number.

User also needs to specify the machine type of SERIES(N1) and N1 Standard-8.

User also needs to modify the boot disk and set the operating system as Ubuntu. The Ubuntu version needs to 18.04 LTS. The size must be set to 20 GB.

In Access scopes, 'Full access to all Cloud APIs' must be given to enable full access to the VM.

In the Firewall section, 'Allow HTTP traffic' must be enabled.

Then click 'Create'.

After creating the VM instance, the Container Registry API must be enabled. Figure 9.3 shows the Container Registry API creation.

After the VM gets created, connect to SSH; Figure 9.4 shows SSH for created VM and execution.

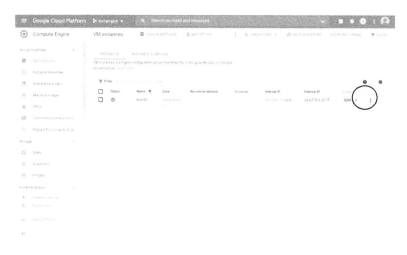

FIGURE 9.3
Container Registry API creation.

```
glms19cs062@text-02:~$ sudo apt install docker.io
Reading package lists... Done
Building dependency tree
Reading state information... Done
The following package was automatically installed and is no longer required:
  libnuma1
Use 'sudo apt autoremove' to remove it.
The following additional packages will be installed:
  bridge-utils containerd pigz runc ubuntu-fan
Suggested packages:
  ifupdown aufs-tools cgroupfs-mount | cgroup-lite debootstrap docker-doc rinse zfs-fuse | zfsutils
The following NEW packages will be installed:
  bridge-utils containerd docker.io pigz runc ubuntu-fan
0 upgraded, 6 newly installed, 0 to remove and 6 not upgraded.
Need to get 74.2 MB of archives.
After this operation, 360 MB of additional disk space will be used.
Do you want to continue? [Y/n] []
```

FIGURE 9.4
SSH created VM and executing the commands.

The VM then opens in a new tab.

Run the command **sudo apt update** to install packages.

Docker is not downloaded by default, so run the command **sudo apt install docker.io**.

Set the PROJECT_ID and the Image URI as variables.

1. The training code should be organized in the following way

```
cloud_training folder:
    -trainer folder:
        -__init__.py
        - glove.6B.300d.txt
        -task.py
        -model.py
    -Dockerfile
    -config.yaml
    -README.md
```

2. task.py contains all the instructions to obtain the dataset, process it, create the model and run the training loop. It also specifies the distribution strategy to be used; in this case it is the Multi Worker Mirrored Strategy.

```
1.   import tensorflow as tf
2.   import tensorflow_datasets as tfds
3.   import argparse
4.   import os
5.   import numpy as np
6.   import pandas as pd
7.   from tensorflow.keras.preprocessing.text import
         Tokenizer
8.   from tensorflow.keras.preprocessing.sequence
         import pad_sequences
9.   from sklearn.model_selection import
         train_test_split
10.  from trainer.model import create_model
11.
12.  PER_REPLICA_BATCH_SIZE = 64
13.
14.  def get_args():
15.      '''Parses args.'''
16.
17.      parser = argparse.ArgumentParser()
18.      parser.add_argument(
19.          '--epochs',
20.          required=True,
21.          type=int,
22.          help='number training epochs')
23.      parser.add_argument(
24.          '--job-dir',
25.          required=True,
```

```
26.            type=str,
27.            help='bucket to save model')
28.      args = parser.parse_args()
29.      return args
30.
31.
32.    def generate_embedding_matrix(glove_file_path,
          w_index, embed_dimension):
33.        '''Generates an embedding matrix for the
             dataset using GloVe: Global Vectors for
             Word Representation'''
34.
35.        size_of_vocab = len(w_index) + 1
36.        embed_mat = np.zeros((size_of_vocab,
             embed_dimension))
37.
38.        with open(glove_file_path,encoding='utf-8')
             as file:
39.          for line in file:
40.              w, *vec = line.split()
41.              if w in w_index:
42.                  index = w_index[w]
43.                  embed_mat[index] = np.array(vec,
                        dtype=np.float32)
                        [:embed_dimension]
44.
45.        return size_of_vocab, embed_dimension,
             embed_mat
46.
47.
48.    def preprocess_data(comment_train,comment_test):
49.        '''Tokensizes the dataset, i.e assigns
             a unique number to each word of the
             dataset'''
50.
51.      tokenizer = Tokenizer()
52.      tokenizer.fit_on_texts(comment_train)
53.
54.      number_of_words=8500
55.      tokenizer.word_index = {x:y for x,y in
             tokenizer.word_index.items() if y <=
             number_of_words}
56.      tokenizer.word_index[tokenizer.oov_token] =
             number_of_words + 1
57.
58.      comment_train = tokenizer.texts_to_sequences
             (comment_train)
```

```
59.     comment_test = tokenizer.
           texts_to_sequences(comment_test)
60.     size_of_vocab = len(tokenizer.word_index) + 1
61.
62.     max_length = 50
63.     comment_train = pad_sequences(comment_train,
           padding='post', maxlen=max_length)
64.     comment_test = pad_sequences(comment_test,
           padding='post', maxlen=max_length)
65.
66.     embed_dimension = 300
67.     size_of_vocab, embed_dimension, embed_mat
           = generate_embedding_matrix('trainer/
           glove.6B.300d.txt', tokenizer.word_index,
           embed_dimension)
68.
69.     return comment_train, comment_test, size_
           of_vocab, embed_dimension, embed_mat,
           max_length
70.
71.
72.  def create_dataset(batch_size):
73.     '''loads a tf dataset, converts to csv and
           creates the training and testing sets '''
74.
75.     initial_data_load = tfds.
           load(name='goemotions', split='train')
76.     initial_data_load = tfds.
           as_dataframe(initial_data_load)
77.     initial_data_load.to_csv("converted_dataset.
           csv")
78.     ds = pd.read_csv('converted_dataset.csv')
79.
80.     comment = ds['comment_text'].values
81.     comment= comment.astype(str)
82.
83.     emo=ds.drop(['comment_text','Unnamed:
           0'],axis=1).values
84.     comment_train, comment_
           test, emo_train, emo_test =
           train_test_split(comment,emo,test_size=0.15)
85.
86.     return comment_train, comment_test, emo_train,
           emo_test
87.
88.
89.  def _is_chief(task_type, task_id):
90.     '''Determines of machine is chief.'''
91.
```

```
92.      return task_type == 'chief'
93.
94.
95.   def _get_temp_dir(dirpath, task_id):
96.      '''Gets temporary directory for saving
             emo_classifier.'''
97.
98.      base_dirpath = 'workertemp_' + str(task_id)
99.      temp_dir = os.path.join(dirpath, base_dirpath)
100.        tf.io.gfile.makedirs(temp_dir)
101.        return temp_dir
102.
103.
104.     def write_filepath(filepath, task_type,
             task_id):
105.        '''Gets filepath to save
              emo_classifier.'''
106.
107.        dirpath = os.path.dirname(filepath)
108.        base = os.path.basename(filepath)
109.        if not _is_chief(task_type, task_id):
110.          dirpath = _get_temp_dir(dirpath, task_id)
111.        return os.path.join(dirpath, base)
112.
113.
114.     def main():
115.        args = get_args()
116.        strategy = tf.distribute.
             MultiWorkerMirroredStrategy()
117.        global_batch_size = PER_REPLICA_BATCH_SIZE
             * strategy.num_replicas_in_sync
118.
119.        comment_train, comment_
             test, emo_train, emo_test =
             create_dataset(global_batch_size)
120.        comment_train, comment_test, size_of_
             vocab, embed_dimension, embed_mat, max_
             length =preprocess_data(comment_train,
             comment_test)
121.
122.        with strategy.scope():
123.          model = create_model(size_of_vocab,
               embed_dimension, embed_mat,
               max_length)
124.
125.        model.fit(comment_train, emo_train,
             epochs=args.epochs, validation_
             data=(comment_test, emo_test))
126.
```

```
127.          # Determine type and task of the machine
                 from the strategy cluster resolver
128.          task_type, task_id = (strategy.cluster_
                 resolver.task_type, strategy.cluster_
                 resolver.task_id)
129.
130.          # Based on the type and task, write to the
                 desired model path
131.          write_model_path = write_filepath(args.
                 job_dir, task_type, task_id)
132.          model.save(write_model_path)
133.
134.     if __name__ == "__main__":
135.         main()
```

3. model.py contains the model architecture that we'll be using. The model can also be specified in task.py, thus eliminating the need for model.py.

```
1. import tensorflow as tf
2.
3. from tensorflow.keras.layers import Dropout
4. from tensorflow.keras.models import Sequential
5. from tensorflow.keras import layers
6.
7.
8. def create_model(size_of_vocab, embed_dimension,
       embed_mat, max_length):
9.
10.   emo_classifier = Sequential()
11.   emo_classifier.add(layers.Embedding(size_of_
          vocab, embed_dimension, weights=[embed_mat],
          input_length=max_length, trainable=True))
12.   emo_classifier.add(layers.Conv1D(256, 3,
          activation='relu'))
13.   emo_classifier.add(Dropout(0.5))
14.   emo_classifier.add(layers.GlobalMaxPooling1D())
15.   emo_classifier.add(layers.Dense(28,
          activation='sigmoid'))
16.   opt = tf.optimizers.Adam(learning_rate=0.0002)
17.   emo_classifier.compile(optimizer=opt,
          loss='binary_crossentropy')
18.
19.   return emo_classifier
20.
```

4. The write_filepath method inside the sample task.py will save the model to the cloud storage bucket automatically once the model has finished training, and is therefore essential to include in the program.

5. config.yaml and its functioning are described in the configuring clusters section.

```
 1. trainingInput:
 2.   scaleTier: CUSTOM
 3.   masterType: n1-standard-8
 4.   masterConfig:
 5.     imageUri: gcr.io/multigpu/text:v1
 6.   useChiefInTfConfig: true
 7.   workerType: n1-standard-8
 8.   workerCount: 1
 9.   workerConfig:
10.     imageUri: gcr.io/multigpu/text:v1
```

Pushing the docker image to the container registry

1. Use the dockerfile provided or write your own to create an image to run the training job from. This dockerfile will specify the base image and install all the other prerequisites to run the training job successfully.

2. Run the following command to name and build the docker image

   2.1. **export PROJECT_ID=$(gcloud config list project --format "value(core.project)")**

   2.2. **export IMAGE_URI=gcr.io/$PROJECT_ID/{repo_name}:{tag}**

```
glms19cs062@text-02:~$ export PROJECT_ID=$(gcloud config list project --form
at "value(core.project)")
glms19cs062@text-02:~$ echo $PROJECT_ID
text-project-02
glms19cs062@text-02:~$ export IMAGE_URI=gcr.io/$PROJECT_ID/multitext:v1
glms19cs062@text-02:~$ echo $IMAGE_URI
gcr.io/text-project-02/multitext:v1
```

Repo name will be the name of your object in the container registry and tag will differentiate it from other similarly named containers.

Example **export IMAGE_URI=gcr.io/$PROJECT_ID/multitext:v1**

2.3. sudo docker build -f Dockerfile -t $IMAGE_URI.

2.4. gcloud auth configure-docker

```
glms19cs062@text-02:~$ gcloud auth configure-docker
Adding credentials for all GCR repositories.
WARNING: A long list of credential helpers may cause delays running 'docker build'. We recommend passing the regi
stry name to configure only the registry you are using.
After update, the following will be written to your Docker config file located at
[/home/glms19cs062/.docker/config.json]:
{
  "credHelpers": {
    "gcr.io": "gcloud",
    "us.gcr.io": "gcloud",
    "eu.gcr.io": "gcloud",
    "asia.gcr.io": "gcloud",
    "staging-k8s.gcr.io": "gcloud",
    "marketplace.gcr.io": "gcloud"
  }
}

Do you want to continue (Y/n)?  Y

Docker configuration file updated.
```

2.5. sudo docker push $IMAGE_URI

After the docker image is finished pushing to GCP, navigate to the container registry section and find the image listed.

```
glms19cs062@text-02:~$ sudo docker push $IMAGE_URI
The push refers to repository [gcr.io/text-project-02/multitext]
a1076cb05624: Pushing [=============>                        ]  239.5MB/1.038GB
5bb1aa5df10d: Pushed
f028010939aa: Pushed
dc99c4ea3a81: Pushed
37b508c5711b: Pushed
756ab564e194: Pushing [=====>                                ]  535.3MB/4.873GB
2ae86808a3d1: Pushed
1dccbdf9b557: Pushed
cfcbdbc2b748: Pushed
937ab8f29c2e: Pushed
5d417b2f7486: Pushed
d6a297a3e6e4: Pushed
6474a5e8117f: Pushed
fe498124ed57: Pushed
d5454704bb3d: Pushed
fb896ef24b4b: Pushing [====================================>]  244.9MB/295.2MB
5087113f67c8: Pushed
2a92857a1d48: Pushed
0ded97864c52: Pushing [====>                                 ]  46.14MB/559.6MB
b50bbaac3e32: Pushing [==>                                   ]  42.44MB/742.5MB
262ea1af4c10: Waiting
b420a468ca49: Waiting
608c205798d1: Waiting
0760cd6d4269: Waiting
fb4755c89c2a: Waiting
22cfb9034da6: Waiting
8bec4fbfce85: Waiting
3b129ca3db46: Waiting
64cb1a1930ab: Preparing
600ef5a43f1f: Preparing
8f8f0266f834: Waiting
```

2.6. sudo docker images

Run sudo docker images to see the active images.

```
glms19cs062@text-02:~$ sudo docker images
REPOSITORY                                        TAG     IMAGE ID       CREATED         SIZE
gcr.io/text-project-02/multitext                  v1      627780a63545   10 minutes ago  16.3GB
gcr.io/deeplearning-platform-release/tf2-gpu.2-5  latest  307b41b1aec7   3 months ago    15.3GB
glms19cs062@text-02:~$
```

9.3.2 Running using Single-Worker/Distributed Instance

The config.yaml file must be updated with the new image URI, and the worker count has to be given as 1.

```
CLOUD SHELL
Terminal    (text-project-02) ×  + ▾                                    ✎ Open Editor

Welcome to Cloud Shell! Type "help" to get started.
Your Cloud Platform project in this session is set to text-project-02.
Use "gcloud config set project [PROJECT_ID]" to change to a different project.
glms19cs062@cloudshell:~ (text-project-02)$ cat config.yaml
trainingInput:
  scaleTier: CUSTOM
  masterType: n1-standard-8
  masterConfig:
    imageUri: gcr.io/text-project-02/multitext:v1
  useChiefInTfConfig: true
  workerType: n1-standard-8
  workerCount: 1
  workerConfig:
    imageUri: gcr.io/text-project-02/multitext:v1
```

1. To launch a training job on gcp, run the following command from the local project directory.

```
1. gcloud ai-platform jobs submit training {job_name} \
2.     --region europe-west2 \
3.     --config config.yaml \
4.     --job-dir gs://{gcs_bucket/model_dir} -- \
5.     --epochs 5 \
6.
```

2. Replace {job_name} with a name for the training job.

3. The --region argument specifies the gcp region the training job would run in.

4. The --job-dir argument specifies which bucket and directory to store the final model in; the gcs_bucker name should be the same as the one initially created in the cloud storage section. The model_dir can be specified as any unique name, this directory will be automatically created in the specified gcp bucket.

5. The --epochs flag specifies the number of epochs for which training will be performed.

6. The training job takes about ten minutes to be queued. The status of the training job and logs can be viewed in the AI Platform/jobs section on the GCP console.

```
q1ms19cs062@cloudshell:~ (text-project-02)$ qcloud ai-platform jobs describe text_model
createTime: '2021-11-15T17:01:28Z'
etag: Uux3o axYbM=
jobid: text_model
jobPosition: '1'
state: QUEUED
trainingInput:
  args:
  - --epochs
  - '5'
  jobDir: gs://text-model/
  masterConfig:
    imageUri: gcr.io/text-project-02/multitext:v1
  masterType: n1-standard-8
  region: europe-west2
  scaleTier: CUSTOM
  useChiefInTfConfig: true
  workerConfig:
    imageUri: gcr.io/text-project-02/multitext:v1
  workerCount: '1'
  workerType: n1-standard-8
trainingOutput: {}

View job in the Cloud Console at:
https://console.cloud.google.com/mlengine/jobs/text_model?project=text-project-02

View logs at:
https://console.cloud.google.com/logs?resource=ml_job%2Fjob_id%2Ftext_model&project=text-project-02
```

Navigate to AI Platform to see the completion of the job. Figure 9.5 shows the same.

9.3.3 Deployment and Prediction

To deploy the model, we need to first import the model to Vertex AI. After the job is completed in AI platform, navigate to Vertex AI and import the model. Figure 9.6 shows different steps in deployment and prediction of model.

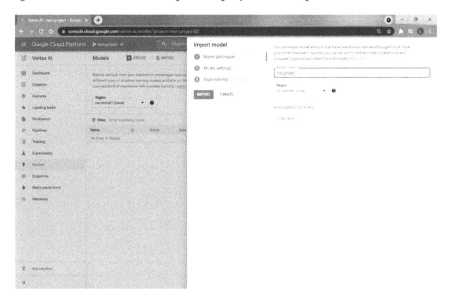

FIGURE 9.5
AI platform to display job.

FIGURE 9.6
Deployment and prediction of model.

Specify the Model Name

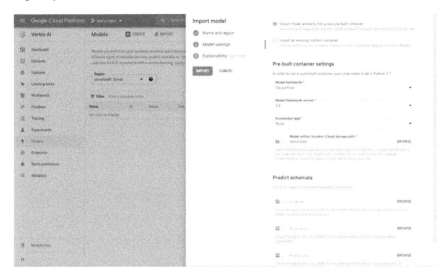

Specify the TensorFlow version and model framework and give the location of the model in the Cloud Bucket.

Then can import the model and model will be ready to add the endpoints (Online Prediction) and perform batch predictions.

Create the endpoints only if the user wants to perform online prediction.

To do batch/online prediction, we must verify that the data used for prediction is the same as the data used to train the model. Because the model is trained with encoded text in the form of a numpy array in this example, the prediction audio must be transformed to a numpy array before being put in a JSON file.

The following script once executed takes a single audio data from test data and saves it in the JSON file.

Open the Cloud shell and ensure that the dataset 'glove.6B.300d.txt' is downloaded in the shell environment. Source (https://bit.ly/32R9WWP)

```
1.  import tensorflow as tf
2.  import tensorflow_datasets as tfds
3.  import os
4.  import numpy as np
5.  import pandas as pd
6.  from tensorflow.keras.preprocessing.text import
        Tokenizer
7.  from tensorflow.keras.preprocessing.sequence
        import pad_sequences
8.  from sklearn.model_selection import
        train_test_split
```

```
9.
10.    import json
11.    PER_REPLICA_BATCH_SIZE = 64
12.
13.    def generate_embedding_matrix(glove_file_path, w_
          index, embed_dimension):
14.        '''Generates an embedding matrix for the
              dataset using GloVe: Global Vectors for
              Word Representation'''
15.
16.        size_of_vocab = len(w_index) + 1
17.        embed_mat = np.zeros((size_of_vocab,
              embed_dimension))
18.
19.        with open(glove_file_path,encoding='utf-8')
              as file:
20.            for line in file:
21.                w, *vec = line.split()
22.                if w in w_index:
23.                    index = w_index[w]
24.                    embed_mat[index] = np.array(vec,
                          dtype=np.float32)
                          [:embed_dimension]
25.
26.        return size_of_vocab, embed_dimension,
              embed_mat
27.
28.    def preprocess_data(comment_train,comment_test):
29.        '''Tokensizes the dataset, i.e assigns a unique
              number to each word of the dataset'''
30.
31.        tokenizer = Tokenizer()
32.        tokenizer.fit_on_texts(comment_train)
33.
34.        number_of_words=8500
35.        tokenizer.word_index = {x:y for x,y in
              tokenizer.word_index.items() if y <=
              number_of_words}
36.        tokenizer.word_index[tokenizer.oov_token] =
              number_of_words + 1
37.
38.        comment_train = tokenizer.texts_to_sequences
              (comment_train)
39.        comment_test = tokenizer.texts_to_sequences
              (comment_test)
40.        size_of_vocab = len(tokenizer.word_index) + 1
41.
42.        max_length = 50
```

```
43.     comment_train = pad_sequences(comment_train,
            padding='post', maxlen=max_length)
44.     comment_test = pad_sequences(comment_test,
            padding='post', maxlen=max_length)
45.
46.     embed_dimension = 300
47.     size_of_vocab, embed_dimension, embed_mat =
            generate_embedding_matrix('glove.6B.300d.
            txt', tokenizer.word_index, embed_dimension)
48.
49.     return comment_train, comment_test, size_
            of_vocab, embed_dimension, embed_mat,
            max_length
50.
51.  def create_dataset(batch_size):
52.     '''loads a tf dataset, converts to csv and
            creates the training and testing sets '''
53.
54.     initial_data_load = tfds.load(name='goemotions',
            split='train')
55.     initial_data_load = tfds.as_dataframe
            (initial_data_load)
56.     initial_data_load.to_csv("converted_dataset.csv")
57.     ds = pd.read_csv('converted_dataset.csv')
58.
59.     comment = ds['comment_text'].values
60.     comment= comment.astype(str)
61.
62.     emo=ds.drop(['comment_text','Unnamed:
            0'],axis=1).values
63.     comment_train, comment_test, emo_train,
            emo_test = train_test_split
            (comment,emo,test_size=0.15)
64.
65.     return comment_train, comment_test, emo_train,
            emo_test
66.
67.  def main():
68.     epochs=5
69.     job_dir="/home/" '''Enter the directory for
            JSON to be saved '''
70.
71.     strategy = tf.distribute.
            MultiWorkerMirroredStrategy()
72.     global_batch_size = PER_REPLICA_BATCH_SIZE *
            strategy.num_replicas_in_sync
73.
74.     comment_train, comment_test, emo_train, emo_
            test = create_dataset(global_batch_size)
```

```
75.     comment_train, comment_test, size_of_
            vocab, embed_dimension, embed_mat, max_
            length =preprocess_data(comment_train,
            comment_test)
76.     lists = comment_test.tolist()
77.     lists=lists[1]
78.     json_str = json.dumps(lists)
79.     name="text_ip.jsonl"
80.     with open(name,'w') as test_file:
81.             test_file.write(json_str)
82.
83.
84. if __name__ == "__main__":
85.     main()
86.
```

Once executed, a JSON file will be seen in the environment (change the job_dir in main()). This json is then used to perform batch prediction.

Then can navigate to Batch Prediction to perform batch predictions. Figure 9.7 shows different steps involved in batch prediction.

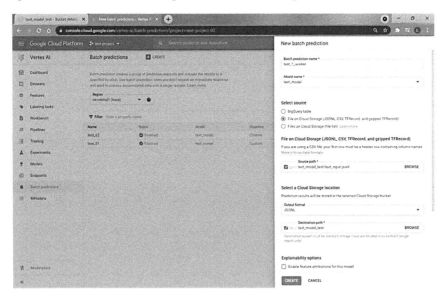

{"instance": [6, 33, 150, 201, 850, 17, 2, 2046, 16, 3116, 13, 1818, 31, 2442, 33, 163, 201, 5, 103, 0, 0, 0, 0, 0, 0, 0, 0, 0, 0, 0, 0, 0, 0, 0, 0, 0, 0, 0, 0, 0, 0, 0, 0, 0, 0, 0, 0, 0, 0, 0, 0, 0, 0, 0, 0, 0, 0, 0, 0, "prediction": [0.108222634, 0.0951084197, 0.0324407816, 0.0786496103, 0.301110089, 0.0784325898, 0.204517573, 0.315691292, 0.0458306874, 0.100201249, 0.210002545, 0.0312336385, 0.0234571397, 0.0586290359, 0.0436365902, 0.0425846577, 0.0331179798, 0.0435993969, 0.0484809279, 0.0302815735, 0.734952688, 0.13656351, 0.0291862786, 0.105991632, 0.0435085787, 0.03428334, 0.0673542321, 0.0712715387]}

FIGURE 9.7
Batch prediction process.

Specify the parameters with the input file and the output format.

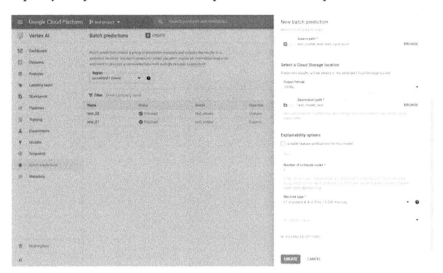

Choose Compute Nodes as 1 for single worker and machine type as n1 standard 4.

After the batch prediction is done, navigate to the export location to see the prediction.

The prediction is as follows.

Next run batch prediction for four workers with the same machine type. Figure 9.8 shows batch prediction process with four workers.

{"instance": [6, 33, 158, 201, 850, 17, 2, 2946, 16, 3116, 23, 1018, 31, 2442, 33, 163, 201, 5, 103, 0, 0, 0, 0, 0, 0, 0, 0, 0, 0, 0, 0, 0, 0, 0, 0, 0, 0, 0, 0, 0, 0, 0, 0, 0, 0, 0, 0, 0, 0, 0, 0, 0, 0], "prediction": [0.108222634, 0.8951064197, 0.0324407816, 0.0786496103, 0.301110089, 0.0784325898, 0.204517573, 0.315691292, 0.0458306074, 0.100201249, 0.210092545, 0.0312336385, 0.0234571397, 0.0586290359, 0.0436365902, 0.0425846577, 0.0331179798, 0.0435993069, 0.0484809279, 0.0302015735, 0.734952688, 0.13656351, 0.0291062786, 0.105991632, 0.0435885787, 0.03428334, 0.0673542321, 0.0712715307]}

FIGURE 9.8
Batch prediction process with four workers.

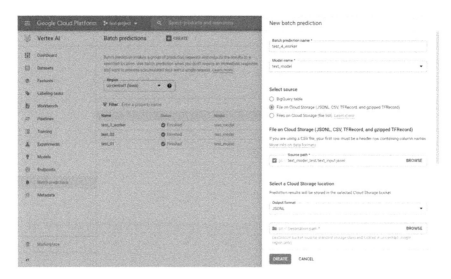

Specify the model name, source path and the output path.

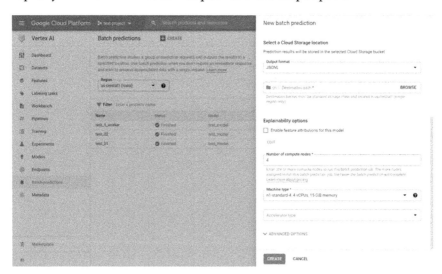

Specify the workers as 4 and machine type as n1 standard 4.
Run the batch prediction and see the following results.

The prediction is as follows

After executing both batches' predictions with the same input, we can observe the following results. Table 9.1 shows the results.

9.3.4 Hyperparameters and Optimization

For hyperparameter, we test the model with the same input file with 1 and 4 workers but change the machine type to n1-cpu4, which has higher CPU power.

Navigate to batch prediction and set the following details. Figure 9.9 shows hyperparameters and optimization for batch prediction.

TABLE 9.1

Batch Processing Output

Instance	Elapsed Time
Single-worker Instance	21 min 14 sec
Distributed Instance (4 instances)	20 min 34 sec

{"instance": [6, 33, 158, 201, 850, 17, 2, 2946, 16, 3116, 23, 1018, 31, 2442, 33, 163, 201, 5, 103, 0, 0, 0, 0, 0, 0, 0, 0, 0, 0, 0, 0, 0, 0, 0, 0, 0, 0, 0, 0, 0, 0, 0, 0, 0, 0, 0, 0, 0], "prediction": [0.108222634, 0.0951084197, 0.022440710, 0.0786496103, 0.301110089, 0.0784325098, 0.204517573, 0.315691292, 0.0458306074, 0.100201249, 0.210002545, 0.0312336385, 0.0234571397, 0.0586290359, 0.0436305902, 0.0425846577, 0.0331179798, 0.0435993969, 0.0484809279, 0.0302815735, 0.734052688, 0.13656351, 0.0291862786, 0.105991632, 0.0435885787, 0.03428334, 0.0673542321, 0.0712715387]}

FIGURE 9.9
Hyperparameters and optimization for batch prediction.

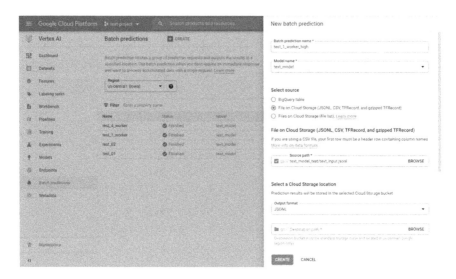

Set the number of workers to 1.

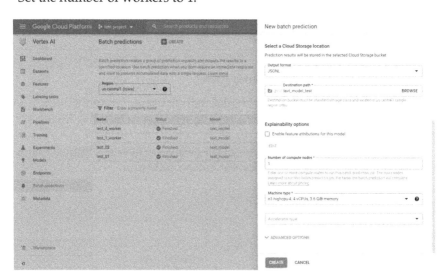

The results are as follows after the prediction.

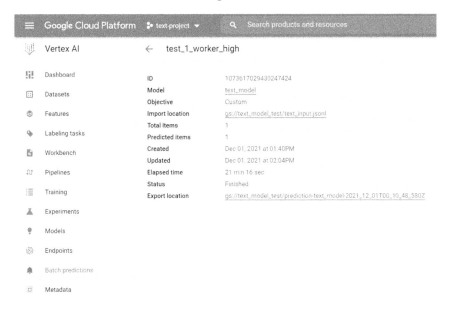

Run the same way for four workers with the higher CPU type. Figure 9.10 shows the process of batch prediction with higher CPU type

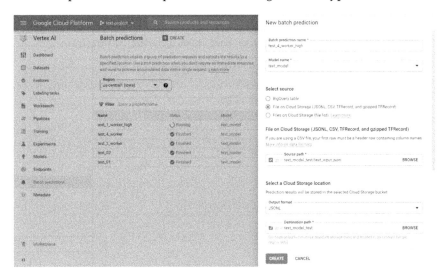

{"instance": [6, 33, 158, 201, 850, 17, 2, 2946, 16, 3116, 23, 1010, 31, 2442, 33, 163, 201, 5, 103, 0, 0, 0, 0, 0, 0, 0, 0, 0, 0, 0, 0, 0, 0, 0, 0, 0, 0, 0, 0, 0, 0, 0, 0, 0, 0, 0, 0], "prediction": [0.108222654, 0.0951084197, 0.0324407816, 0.0786496103, 0.301110089, 0.0784325898, 0.204517573, 0.315691292, 0.0458306074, 0.100201249, 0.210092545, 0.0312136385, 0.0234571397, 0.0586290359, 0.0436365902, 0.0425846577, 0.0331179798, 0.0435993969, 0.0484809279, 0.0302815735, 0.734952688, 0.13656351, 0.0291862786, 0.105991632, 0.0435885787, 0.03428334, 0.0673542321, 0.0712715387]}

FIGURE 9.10
Batch prediction with higher CPU type.

Set the worker count to 4.

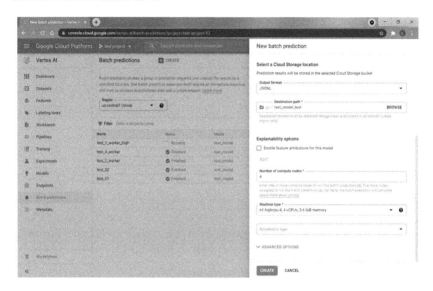

After the prediction is completed, find the following prediction.

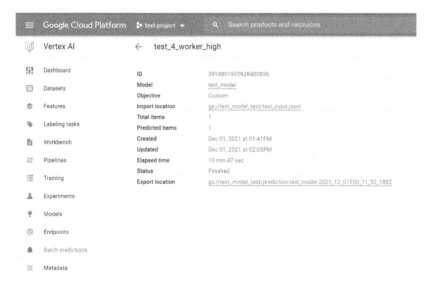

Navigate to the export location.

References

1. Moreno, A., & Redondo, T. (2016). Text analytics: The convergence of big data and artificial intelligence. *IJIMAI*, 3(6), 57–64.
2. Galati, F., & Bigliardi, B. (2019). Industry 4.0: Emerging themes and future research avenues using a text mining approach. *Computers in Industry*, 109, 100–113.
3. Maas, A., Daly, R. E., Pham, P. T., Huang, D., Ng, A. Y., & Potts, C. (2011, June). Learning word vectors for sentiment analysis. In *Proceedings of the 49th annual meeting of the association for computational linguistics: Human language technologies* (pp. 142–150).

Exercises

1. Under which division can you change the operating system and boot disk type?
2. Which option must be checked to allow all the Cloud APIs for a VM?
3. Which command is used to download docker in a VM instance?
4. Which command must be used to store variables in the Cloud shell?
5. What are the parameters to import the model?

10

Speech Analytics

10.1 Introduction

Speech analytics is an upcoming research area because of its applications like e-learning, entertainment, computer games and surveillance systems. However, there exists a significantly challenging task of using the right learning techniques for detection of the emotion. The reasons or the drawbacks of the uncertainty of the learning are due to the noise and not choosing the right technique [1]. Artificial intelligence (AI) could have a potential impact on building speech emotion applications with human–computer interactions [2]. There has been considerable effort in this area to develop more AI applications.

  Deep learning models can be employed for such recognition tasks [3]. Various deep learning models are used for speech emotion recognition and fall into two groups. One of the group objectives is detecting significant features from the raw samples, while the other group's focus is on the representation of the sound in the input files [4, 5]. Most of these models have achieved excellent results using deep belief networks (DBNs), convolutional neural networks (CNNs) and long short-term memory (LSTM). Gated recurrent units are least explored in the speech emotion recognition applications. These neural networks provide the advantage of storing the information from the previous state and passing it to the next state for prediction purposes.

10.2 Data Preparation with Google Cloud

Speech analytics, also known as interaction analytics, is a type of artificial intelligence-based technology that understands, processes and analyzes human speech. Speech analytics is used in various sectors like contact centers to evaluate call recordings and transcripts from digital channels like chat and text messages. In this chapter, we will be utilizing a dataset on speech commands [6] consisting of audio files in the WAV format, which can be used to build and train a basic automatic speech recognition (ASR) model for recognizing eight different words.

10.3 Speech Classification with Google Cloud and TensorFlow

10.3.1 Create a TensorFlow Application

We use the TensorFlow built-in dataset to run this model.

As this model needs a lot of memory, it is needed to create a VM (virtual machine) instance, rather than using the cloud shell, which would run out of disk space.

For creating the VM instance, user needs to enable the Compute Engine API.

On the navigation bar, navigate to APIs & Services and enable the Compute Engine API. Figure 10.1 shows the steps to enable Compute Engine API.

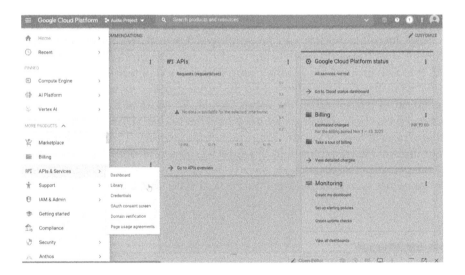

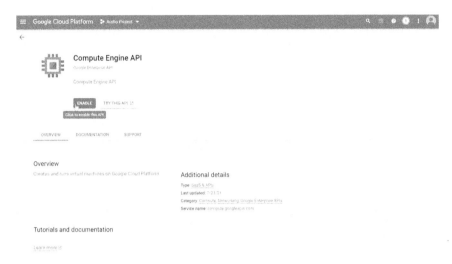

FIGURE 10.1
Enabling the Compute Engine API.

After enabling the API, navigate to Compute Engine/VM Instances, Figure 10.2 shows the steps to configure Compute Engine instance.

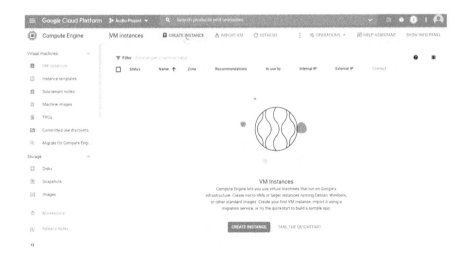

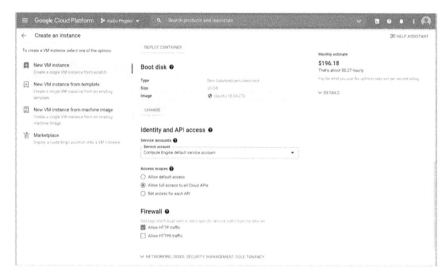

FIGURE 10.2
Configuring Compute Engine instance.

User needs to specify the instance name ending with a number.

User also needs to specify the machine type of SERIES(N1) and N1 Standard-8.

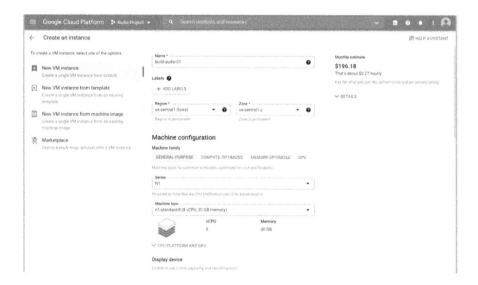

User also needs to modify the boot disk and set the operating system as Ubuntu. The Ubuntu version needs to be 18.04 LTS. The size must be set to 20 GB.

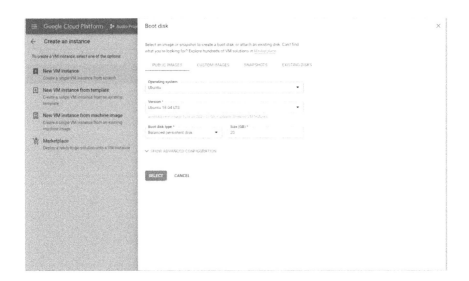

In Access scopes, full access to all Cloud APIs must be given to enable full access to the VM.

In the Firewall section, allow HTTP traffic must be enabled.

Then click 'Create'.

After creating the VM instance, the Container Registry API must be enabled.

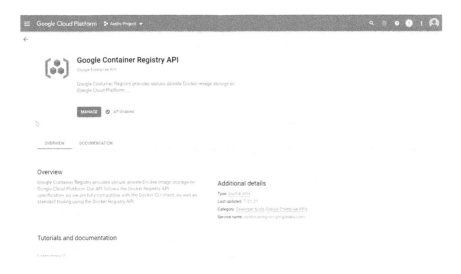

After the VM gets created, connect to SSH. Figure 10.3 shows steps to connect to VM and run commands.

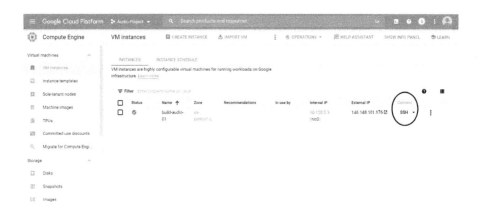

The VM then opens in a new tab.

```
glms19cs062@build-audio-01:~$ sudo docker build -f Dockerfile -t $IMAGE_URI .
Sending build context to Docker daemon  48.13kB
Step 1/4 : FROM gcr.io/deeplearning-platform-release/tf2-gpu.2-5
 ---> 307b41b1aec7
Step 2/4 : WORKDIR /root
 ---> Using cache
 ---> 9d3283b65a3d
Step 3/4 : COPY trainer/ /root/trainer/
 ---> Using cache
 ---> dc7a50cb3e83
Step 4/4 : ENTRYPOINT ["python", "-m", "trainer.task"]
 ---> Using cache
 ---> 0a4cc55b1a83
Successfully built 0a4cc55b1a83
Successfully tagged gcr.io/audio-project-01/multicore:v1
```

FIGURE 10.3
Connect to VM and run commands.

Run the command **sudo apt update** to install packages.

Docker is not downloaded by default so run the command **sudo apt install docker.io**.

```
glms19cs062@build-audio-01:~$ sudo apt install docker.io
Reading package lists... Done
Building dependency tree
Reading state information... Done
docker.io is already the newest version (20.10.7-0ubuntu5~18.04.3).
The following package was automatically installed and is no longer required:
  libnuma1
Use 'sudo apt autoremove' to remove it.
0 upgraded, 0 newly installed, 0 to remove and 6 not upgraded.
```

Set the PROJECT_ID and the Image URI as variables.

1. The training code should be organized in the following way: cloud_training folder:

```
-trainer folder:
      -__init__.py
      -task.py
      -model.py
-Dockerfile
-config.yaml
```

2. task.py contains all the instructions to obtain the dataset, process it, create the model and run the training loop. It also specifies the distribution strategy to be used; in this case, it is the Multi Worker Mirrored Strategy.

```
1.  import tensorflow as tf
2.  from tensorflow.keras.layers.experimental import
        preprocessing
3.  from tensorflow.keras import layers
4.  from tensorflow.keras import models
5.  import numpy as np
6.  import argparse
7.  import json
8.  import os
9.  import pathlib
10. import matplotlib.pyplot as plt
11.
12. PER_REPLICA_BATCH_SIZE = 64
13. AUTOTUNE = tf.data.AUTOTUNE
14. commands=np.array(['no','down','left','stop','yes
        ','up','right','go'])
15. def get_args():
16.   '''Parses args.'''
17.
18.   parser = argparse.ArgumentParser()
19.   parser.add_argument(
20.       '--epochs',
21.       required=True,
22.       type=int,
23.       help='number training epochs')
24.   parser.add_argument(
25.       '--job-dir',
26.       required=True,
27.       type=str,
28.       help='bucket to save model')
29.   args = parser.parse_args()
30.   return args
31.
32.
33. def decode_audio(audio_binary):
34.   audio, _ = tf.audio.decode_wav(audio_binary)
35.   return tf.squeeze(audio, axis=-1)
36.
37. def get_label(file_path):
38.   parts = tf.strings.split(file_path, os.path.
        sep)
39.
40.   # Note: You'll use indexing here instead of
        tuple unpacking to enable this
41.   # to work in a TensorFlow graph.
42.   return parts[-2]
43.
44. def get_waveform_and_label(file_path):
```

```
45.     label = get_label(file_path)
46.     audio_binary = tf.io.read_file(file_path)
47.     waveform = decode_audio(audio_binary)
48.     return waveform, label
49.
50.
51.  def get_spectrogram(waveform):
52.     # Padding for files with less than 16000
            samples
53.     zero_padding = tf.zeros([16000] -
            tf.shape(waveform), dtype=tf.float32)
54.
55.     # Concatenate audio with padding so that all
            audio clips will be of the
56.     # same length
57.     waveform = tf.cast(waveform, tf.float32)
58.     equal_length = tf.concat([waveform, zero_
            padding], 0)
59.     spectrogram = tf.signal.stft(
60.         equal_length, frame_length=255,
                frame_step=128)
61.
62.     spectrogram = tf.abs(spectrogram)
63.
64.     return spectrogram
65.
66.
67.  def get_spectrogram_and_label_id(audio, label):
68.     spectrogram = get_spectrogram(audio)
69.     spectrogram = tf.expand_dims(spectrogram, -1)
70.     label_id = tf.argmax(label == commands)
71.     return spectrogram, label_id
72.
73.
74.  def preprocess_dataset(files):
75.     files_ds = tf.data.Dataset.from_tensor_slices
            (files)
76.     output_ds = files_ds.map(get_waveform_and_
            label, num_parallel_calls=AUTOTUNE)
77.     output_ds = output_ds.map(
78.         get_spectrogram_and_label_id,
                num_parallel_calls=AUTOTUNE)
79.     return output_ds
80.
81.
82.  def create_dataset(batch_size):
83.     '''Loads Cassava dataset and preprocesses data.'''
84.
```

```
85.     data_dir = pathlib.Path('data/
            mini_speech_commands')
86.     if not data_dir.exists():
87.       tf.keras.utils.get_file(
88.           'mini_speech_commands.zip',
89.           origin="http://storage.googleapis.com/
                download.tensorflow.org/data/mini_
                speech_commands.zip",
90.           extract=True,
91.           cache_dir='.', cache_subdir='data')
92.
93.     commands = np.array(tf.io.gfile.
            listdir(str(data_dir)))
94.     commands = commands[commands != 'README.md']
95.     print('Commands:', commands)
96.
97.     filenames = tf.io.gfile.glob(str(data_dir) +
            '/*/*')
98.     filenames = tf.random.shuffle(filenames)
99.     num_samples = len(filenames)
100.    print('Number of total examples:',
            num_samples)
101.    print('Number of examples per label:',
102.        len(tf.io.gfile.listdir(str(data_dir/
                commands[0]))))
103.    print('Example file tensor:', filenames[0])
104.
105.    train_files = filenames[:6400]
106.    val_files = filenames[6400: 6400 + 800]
107.    test_files = filenames[-800:]
108.
109.    AUTOTUNE = tf.data.AUTOTUNE
110.    files_ds = tf.data.Dataset.
            from_tensor_slices(train_files)
111.    waveform_ds = files_ds.map(get_waveform_and_
            label, num_parallel_calls=AUTOTUNE)
112.    spectrogram_ds = waveform_ds.map(get_
            spectrogram_and_label_id,
            num_parallel_calls=AUTOTUNE)
113.
114.
115.    train_ds = spectrogram_ds
116.    val_ds = preprocess_dataset(val_files)
117.    test_ds = preprocess_dataset(test_files)
118.
119.
120.    batch_size = 64
121.    train_ds = train_ds.batch(batch_size)
122.    val_ds = val_ds.batch(batch_size)
```

```
123.
124.    train_ds = train_ds.cache().prefetch(AUTOTUNE)
125.    val_ds = val_ds.cache().prefetch(AUTOTUNE)
126.
127.
128.    return train_ds, val_ds, spectrogram_ds
129.
130.
131. def _is_chief(task_type, task_id):
132.    '''Determines of machine is chief.'''
133.
134.    return task_type == 'chief'
135.
136.
137. def _get_temp_dir(dirpath, task_id):
138.    '''Gets temporary directory for saving
           model.'''
139.
140.    base_dirpath = 'workertemp_' + str(task_id)
141.    temp_dir = os.path.join(dirpath, base_dirpath)
142.    tf.io.gfile.makedirs(temp_dir)
143.    return temp_dir
144.
145.
146. def write_filepath(filepath, task_type,
         task_id):
147.    '''Gets filepath to save model.'''
148.
149.    dirpath = os.path.dirname(filepath)
150.    base = os.path.basename(filepath)
151.    if not _is_chief(task_type, task_id):
152.      dirpath = _get_temp_dir(dirpath, task_id)
153.    return os.path.join(dirpath, base)
154.
155.
156. def create_model(spectrogram_ds):
157.    for spectrogram, _ in spectrogram_
           ds.take(1):
158.      input_shape = spectrogram.shape
159.    print('Input shape:', input_shape)
160.    num_labels = len(commands)
161.
162.    norm_layer = preprocessing.Normalization()
163.    norm_layer.adapt(spectrogram_ds.map(lambda x,
           _: x))
164.
165.    model = models.Sequential([
166.      layers.Input(shape=input_shape),
167.      preprocessing.Resizing(32, 32),
```

```
168.        norm_layer,
169.        layers.Conv2D(32, 3, activation='relu'),
170.        layers.Conv2D(64, 3, activation='relu'),
171.        layers.MaxPooling2D(),
172.        layers.Dropout(0.25),
173.        layers.Flatten(),
174.        layers.Dense(128, activation='relu'),
175.        layers.Dropout(0.5),
176.        layers.Dense(num_labels),
177.    ])
178.    model.compile(
179.      optimizer=tf.keras.optimizers.Adam(),
180.      loss=tf.keras.losses.SparseCategoricalCrosse
             ntropy(from_logits=True),
181.      metrics=['accuracy'],
182.    )
183.    return model
184.
185.
186. def main():
187.    args = get_args()
188.    strategy = tf.distribute.
             MultiWorkerMirroredStrategy()
189.    global_batch_size = PER_REPLICA_BATCH_SIZE *
             strategy.num_replicas_in_sync
190.    train_ds,val_ds,spectrogram_ds= create_dataset
             (global_batch_size)
191.    with strategy.scope():
192.      model=create_model(spectrogram_ds)
193.    model.fit(
194.      train_ds,
195.      validation_data=val_ds,
196.      epochs=args.epochs,
197.      callbacks=tf.keras.callbacks.
             EarlyStopping(verbose=1, patience=2),
198.    )
199.
200.    task_type, task_id = (strategy.cluster_
             resolver.task_type,
201.          strategy.cluster_resolver.task_id)
202.
203.    # Based on the type and task, write to the
             desired model path
204.    write_model_path = write_filepath(args.job_
             dir, task_type, task_id)
205.    model.save(write_model_path)
206.
```

```
207.   if __name__ == "__main__":
208.        main()
209.
```

3. model.py contains the model architecture that we'll be using. The model can also be specified in task.py, thus eliminating the need for model.py.

```
1.   import tensorflow as tf
2.   def create_model(number_of_classes):
3.     base_model = tf.keras.applications.
          ResNet50(weights='imagenet',
          include_top=False)
4.     x = base_model.output
5.     x = tf.keras.layers.GlobalAveragePooling2D()(x)
6.     x = tf.keras.layers.Dense(1016, activation=
          'relu')(x)
7.     predictions = tf.keras.layers.Dense(number_of_
          classes, activation='softmax')(x)
8.     model = tf.keras.Model(inputs=base_model.
          input, outputs=predictions)
9.     model.compile(
10.         loss='sparse_categorical_crossentropy',
11.         optimizer=tf.keras.optimizers.Adam(0.0001),
12.         metrics=['accuracy'])
13.    return model
14.
```

4. The write_filepath method inside the sample task.py will save the model to the cloud storage bucket automatically once the model has finished training, and is therefore essential to the program.

5. config.yaml and its functioning is described in the configuring clusters section.

```
1.   trainingInput:
2.     scaleTier: CUSTOM
3.     masterType: n1-standard-8
4.     masterConfig:
5.        imageUri: gcr.io/multigpu/audio:v1
6.     useChiefInTfConfig: true
7.     workerType: n1-standard-8
8.     workerCount: 1
9.     workerConfig:
10.       imageUri: gcr.io/multigpu/audio:v1
11.
```

Pushing the docker image to the container registry

1. Use the dockerfile provided or write your own to create an image to run the training job from. This dockerfile will specify the base image and install all the other prerequisites to run the training job successfully.

```
1.  # Specifies base image and tag
2.  FROM gcr.io/deeplearning-platform-release/
       tf2-gpu.2-5
3.  WORKDIR /root
4.
5.  # Copies the trainer code to the docker image.
6.  COPY trainer/ /root/trainer/
7.
8.  # Sets up the entry point to invoke the trainer.
9.  ENTRYPOINT ["python", "-m", "trainer.task"]
10.
```

2. Run the following command to name and build the docker image.

   2.1. **export PROJECT_ID=$(gcloud config list project --format "value(core.project)")**

   2.2. **export IMAGE_URI=gcr.io/$PROJECT_ID/{repo_name}:{tag}**

Repo name will be the name of your object in the container registry and tag will differentiate it from other similarly named containers.

Example **export IMAGE_URI=gcr.io/$PROJECT_ID/multicore:v1**

   2.3. sudo docker build -f Dockerfile -t $IMAGE_URI.

2.4. gcloud auth configure-docker

```
glms19cs062@build-audio-01:~$ gcloud auth configure-docker
WARNING: Your config file at [/home/glms19cs062/.docker/config.json] contains these credential helper entries:

{
  "credHelpers": {
    "gcr.io": "gcloud",
    "us.gcr.io": "gcloud",
    "eu.gcr.io": "gcloud",
    "asia.gcr.io": "gcloud",
    "staging-k8s.gcr.io": "gcloud",
    "marketplace.gcr.io": "gcloud"
  }
}
Adding credentials for all GCR repositories.
WARNING: A long list of credential helpers may cause delays running 'docker build'. We recommend passing the regi
stry name to configure only the registry you are using.
gcloud credential helpers already registered correctly.
```

2.5. sudo docker push $IMAGE_URI

After the docker image is finished pushing to GCP, navigate to the container registry section and find the image listed.

```
glms19cs062@build-audio-01:~$ sudo docker push $IMAGE_URI
The push refers to repository [gcr.io/audio-project-01/multiaudio]
4cedb34e2c53: Pushed
5bb1aa5df10d: Mounted from deeplearning-platform-release/tf2-gpu.2-5
f028010939aa: Mounted from deeplearning-platform-release/tf2-gpu.2-5
dc99c4ea3a81: Mounted from deeplearning-platform-release/tf2-gpu.2-5
37b508c5711b: Mounted from deeplearning-platform-release/tf2-gpu.2-5
756ab564e194: Mounted from deeplearning-platform-release/tf2-gpu.2-5
2ae86808a3d1: Mounted from deeplearning-platform-release/tf2-gpu.2-5
1dccbdf9b557: Mounted from deeplearning-platform-release/tf2-gpu.2-5
cfcbdbc2b748: Mounted from deeplearning-platform-release/tf2-gpu.2-5
937ab8f29c2e: Mounted from deeplearning-platform-release/tf2-gpu.2-5
5d417b2f7486: Mounted from deeplearning-platform-release/tf2-gpu.2-5
d6a297a3e6e4: Mounted from deeplearning-platform-release/tf2-gpu.2-5
6474a5e8117f: Mounted from deeplearning-platform-release/tf2-gpu.2-5
fe498124ed57: Mounted from deeplearning-platform-release/tf2-gpu.2-5
d5454704bb3d: Mounted from deeplearning-platform-release/tf2-gpu.2-5
fb896ef24b4b: Mounted from deeplearning-platform-release/tf2-gpu.2-5
5087113f67c8: Mounted from deeplearning-platform-release/tf2-gpu.2-5
2a92857a1d48: Mounted from deeplearning-platform-release/tf2-gpu.2-5
0ded97864c52: Mounted from deeplearning-platform-release/tf2-gpu.2-5
b50bbaac3e32: Mounted from deeplearning-platform-release/tf2-gpu.2-5
262ea1af4c10: Mounted from deeplearning-platform-release/tf2-gpu.2-5
b420a468ca49: Mounted from deeplearning-platform-release/tf2-gpu.2-5
608c205798d1: Mounted from deeplearning-platform-release/tf2-gpu.2-5
0760cd6d4269: Mounted from deeplearning-platform-release/tf2-gpu.2-5
fb4755c89c2a: Mounted from deeplearning-platform-release/tf2-gpu.2-5
22cfb9034da6: Mounted from deeplearning-platform-release/tf2-gpu.2-5
8bec4fbfce85: Mounted from deeplearning-platform-release/tf2-gpu.2-5
3b129ca3db46: Mounted from deeplearning-platform-release/tf2-gpu.2-5
64cb1a1930ab: Mounted from deeplearning-platform-release/tf2-gpu.2-5
600ef5a43f1f: Mounted from deeplearning-platform-release/tf2-gpu.2-5
8f8f0266f834: Layer already exists
v1: digest: sha256:fd3668eb2b9cae91b8d0df39a39310bfd3e46623177fe5a93413c27e17227806 size: 6836
```

10.3.2 Running using Single-Worker/Distributed Instance

The config.yaml file must be updated with the new image URI, and the worker count has to be given as 1.

```
g1ms19cs062@cloudshell:~ (audio-project-01)$ cat config.yaml
trainingInput:
  scaleTier: CUSTOM
  masterType: n1-standard-8
  masterConfig:
    imageUri: gcr.io/audio-project-01/multiaudio:v1
  useChiefInTfConfig: true
  workerType: n1-standard-8
  workerCount: 1
  workerConfig:
    imageUri: gcr.io/audio-project-01/multiaudio:v1
```

1. To launch a training job on GCP, run the following command from the local project directory.

```
1.   gcloud ai-platform jobs submit training {job_
        name} \
2.     --region europe-west2 \
3.     --config config.yaml \
4.     --job-dir gs://{gcs_bucket/model_dir} -- \
5.     --epochs 5 \
```

2. Replace {job_name} with a name for the training job.

3. The --region argument specifies the GCP region the training job would run in.

4. The --job-dir argument specifies which bucket and directory to store the final model in; the gcs_bucker name should be the same as the one initially created in the cloud storage section. The model_dir can be specified as any unique name; this directory will be automatically created in the specified GCP bucket.

5. The --epochs flag specifies the number of epochs for which training will be performed.

6. The training job takes about ten minutes to be queued. The status of the training job and logs can be viewed in the AI Platform/jobs section on the GCP console.

10.3.3 Deployment and Prediction

To deploy the model, we need to first import the model to Vertex AI. After the job is completed in AI platform, navigate to Vertex AI and import the model. Figure 10.4 shows the steps for deploying the model.

Specify the model name.

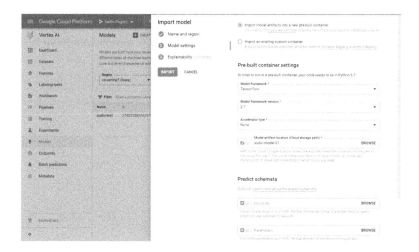

Specify the TensorFlow version and model framework, and give the location of the model in the Cloud Bucket.

Then, import the model, and the model will be ready to add the endpoints (online prediction) and perform batch predictions.

Create the endpoints only if the user wants to perform online prediction.

To do batch/online prediction, we must verify that the data used for prediction is the same as the data used to train the model. Because the model is trained with audio in the form of a spectrogram decoded numpy array in this example, the prediction audio must be transformed to a numpy array before being put in a JSON file.

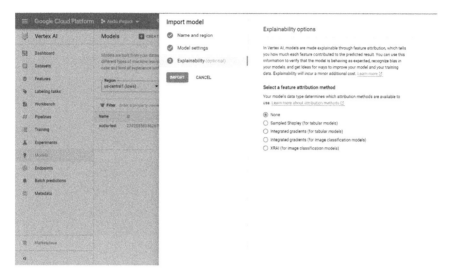

FIGURE 10.4
Deployment of the model.

The following script once executed takes a single audio data from test data and saves it in the JSON file.

```
1.   import tensorflow as tf
2.   from tensorflow.keras.layers.experimental import
        preprocessing
3.   import numpy as np
4.   import json
5.   import os
6.
7.   PER_REPLICA_BATCH_SIZE = 64
8.   AUTOTUNE = tf.data.AUTOTUNE
9.   commands=np.array(['no','down','left','stop','yes
        ','up','right','go'])
10.
11.  def decode_audio(audio_binary):
12.      audio, _ = tf.audio.decode_wav(audio_binary)
13.      return tf.squeeze(audio, axis=-1)
14.
15.  def get_label(file_path):
16.      parts = tf.strings.split(file_path, os.path.
            sep)
17.      # Note: You'll use indexing here instead of
            tuple unpacking to enable this
18.      # to work in a TensorFlow graph.
19.      return parts[-2]
```

```
20.
21.  def get_waveform_and_label(file_path):
22.      label = get_label(file_path)
23.      audio_binary = tf.io.read_file(file_path)
24.      waveform = decode_audio(audio_binary)
25.      return waveform, label
26.
27.  def get_spectrogram(waveform):
28.      # Padding for files with less than 16000
             samples
29.      zero_padding = tf.zeros([16000] -
             tf.shape(waveform), dtype=tf.float32)
30.      # Concatenate audio with padding so that all
             audio clips will be of the
31.      # same length
32.      waveform = tf.cast(waveform, tf.float32)
33.      equal_length = tf.concat([waveform, zero_
             padding], 0)
34.      spectrogram = tf.signal.stft(
35.          equal_length, frame_length=255,
                 frame_step=128)
36.
37.      spectrogram = tf.abs(spectrogram)
38.
39.      return spectrogram
40.
41.  def get_spectrogram_and_label_id(audio, label):
42.      spectrogram = get_spectrogram(audio)
43.      spectrogram = tf.expand_dims(spectrogram, -1)
44.      label_id = tf.argmax(label == commands)
45.      return spectrogram, label_id
46.
47.  def preprocess_dataset(files):
48.      files_ds = tf.data.Dataset.from_tensor_slices
             (files)
49.      output_ds = files_ds.map(get_waveform_and_
             label, num_parallel_calls=AUTOTUNE)
50.      output_ds = output_ds.map(
51.          get_spectrogram_and_label_id,
                 num_parallel_calls=AUTOTUNE)
52.      return output_ds
53.
54.  def create_dataset(batch_size):
55.      '''Loads Cassava dataset and preprocesses
             data.'''
56.
57.      data_dir = pathlib.Path('data/mini_speech_
             commands')
```

```
58.        if not data_dir.exists():
59.            tf.keras.utils.get_file(
60.                'mini_speech_commands.zip',
61.                origin="http://storage.googleapis.com/
                        download.tensorflow.org/data/mini_
                        speech_commands.zip",
62.                extract=True,
63.                cache_dir='.', cache_subdir='data')
64.        filenames = tf.io.gfile.glob(str(data_dir) +
                '/*/*')
65.        filenames = tf.random.shuffle(filenames)
66.        num_samples = len(filenames)
67.        val_files = filenames[6400: 6400 + 800]
68.        val_ds = preprocess_dataset(val_files)
69.        batch_size = 64
70.        val_ds = val_ds.batch(batch_size)
71.        val_ds = val_ds.cache().prefetch(AUTOTUNE)
72.        return  val_ds
73.
74.    def main():
75.        epochs=5
76.        job_dir="/content/" #change the directory to
                save the JSON file
77.        strategy = tf.distribute.
                MultiWorkerMirroredStrategy()
78.        global_batch_size = PER_REPLICA_BATCH_SIZE *
                strategy.num_replicas_in_sync
79.        val_ds= create_dataset(global_batch_size)
80.
81.        for string_, int_ in val_ds.take(1):
82.            # print(string_)
83.            imm = string_.numpy()
84.            lists = imm.tolist()
85.            lists = lists[1]
86.            json_str = json.dumps(lists)
87.            # print(json_str)
88.            name="audio.jsonl"
89.            with open(name,'w') as test_file:
90.                test_file.write(json_str)
91.            print("Saved Input Image in:"+name)
92.
93.    if __name__ == "__main__":
94.        main()
```

This script when executed will provide a JSONL file in your shell environment (change the job_dir in main()). Download the JSONL file and upload it to cloud storage bucket. This JSONL file will be used as the input for the batch prediction.

Then, navigate to Batch Prediction to perform batch predictions in Vertex AI. Figure 10.5 shows different steps for batch prediction.

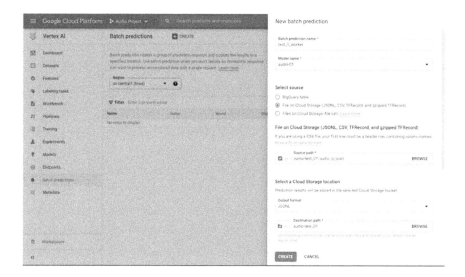

Specify the parameters with the input file and the output format.

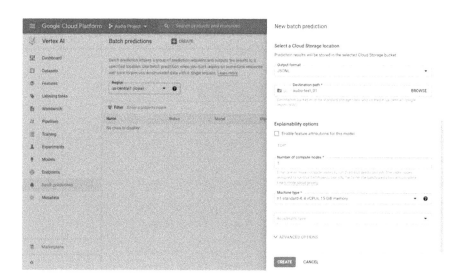

Choose compute nodes as 1 for single worker and machine type as n1 standard 4.

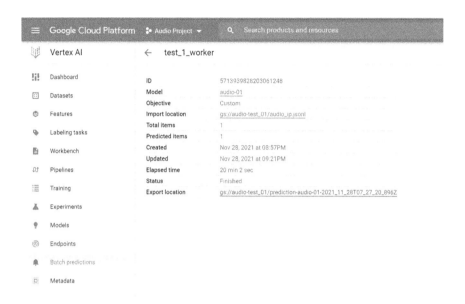

After the batch prediction is done, navigate to the export location to see the prediction.

The prediction is as follows.

```
"prediction": [-0.352441519, -1.93157601, 1.64075649, -0.271354556, 0.780511081, 0.335427731, -1.96720994, -0.527442098]}
```

Next batch prediction for four workers

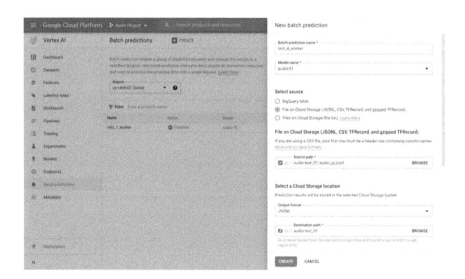

Specify the model name, source path and the output path.

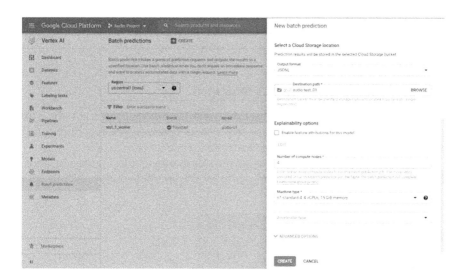

Specify the workers as 4 and machine type as n1 standard 4.
Run the batch prediction and see the following results.

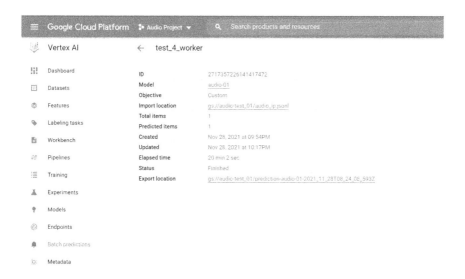

The prediction is as follows

[3.2825369999045506e-05], [8.6079620814416556e-05], [0.00011329806147841737], [0.00014704465866008867]]], "prediction": [-0.352441519, -1.93157601, 1.64075649, -0.271354556, 0.780511081, 0.335427731, -1.96720994, -0.527442008]}

FIGURE 10.5
Batch predictions in Vertex AI.

10.3.3.1 Hyperparameters and Optimization

For hyperparameter, we test the model with the same input file with 1 and 4 workers but change the machine type to n1-cpu4, which has higher CPU power.

Navigate to batch prediction and set the following details. Figure 10.6 shows different steps in batch prediction using hyperparameters and optimization.

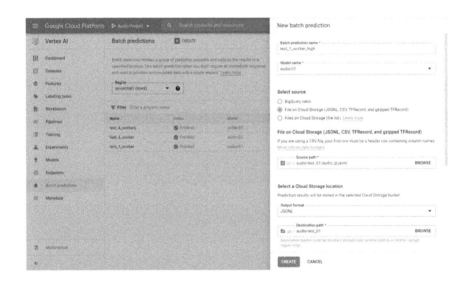

Set the number of workers to 1.

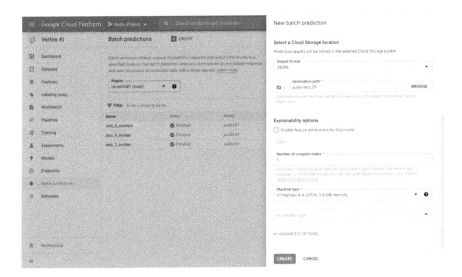

The results are as follows.

The prediction is as follows.

[3.2825369999045506e-05], [8.607962081441656e-05], [0.00011329806147841737], [0.00014704465866088867]]], "prediction": [-0.352441519, -1.93157601, 1.64075649, -0.271354556, 0.780511081, 0.335427731, -1.96720994, -0.527442098])

FIGURE 10.6
Batch prediction using hyperparameters and optimization.

Run the same way for four workers with the higher CPU type. Figure 10.7 shows steps in batch prediction with higher CPU types.

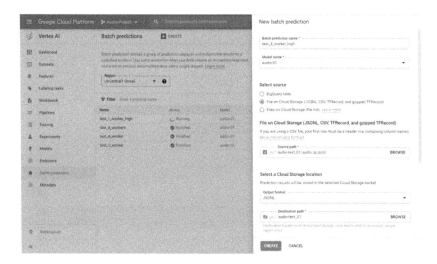

Set the worker count at 4.

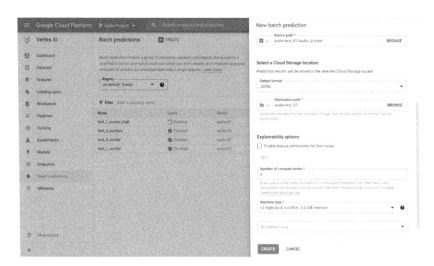

After the prediction is completed, find the following prediction.

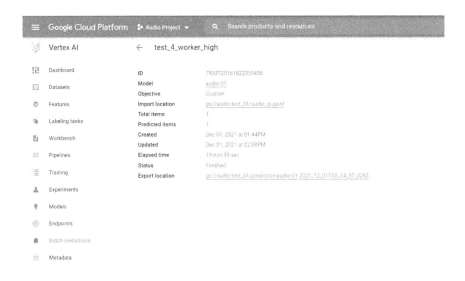

Navigate to the export location.

[8.102746505755931e-05], [8.1347156082113e-05], [3.2825369999045506e-05], [8.607962081441656e-05], [0.0001132980614784137], [0.0001470446586608867]]], "prediction": [-0.352441519, -1.93157601, 1.64075649, -0.271354556, 0.780511081, 0.335427731, -1.96720994, -0.527442098]})

FIGURE 10.7
Batch prediction with higher CPU types.

References

1. Akçay, M. B., & Oğuz, K. (2020). Speech emotion recognition: Emotional models, databases, features, preprocessing methods, supporting modalities, and classifiers. *Speech Communication*, 116, 56–76.
2. Issa, D., Demirci, M. F., & Yazici, A. (2020). Speech emotion recognition with deep convolutional neural networks. *Biomedical Signal Processing and Control*, 59, 101894.
3. Tzirakis, P., Zhang, J., & Schuller, B. W. (2018, April). End-to-end speech emotion recognition using deep neural networks. In *2018 IEEE International Conference on Acoustics, Speech and Signal Processing (ICASSP)* (pp. 5089–5093). IEEE.
4. Huang, C. W., & Narayanan, S. S. (2017, July). Deep convolutional recurrent neural network with attention mechanism for robust speech emotion recognition. In *2017 IEEE international conference on multimedia and expo (ICME)* (pp. 583–588). IEEE.

5. Huang, Y., Tian, K., Wu, A., & Zhang, G. (2019). Feature fusion methods research based on deep belief networks for speech emotion recognition under noise condition. *Journal of Ambient Intelligence and Humanized Computing*, 10(5), 1787–1798.
6. http://storage.googleapis.com/download.tensorflow.org/data/mini_speech_commands.zip

Exercises

1. What is the use of task.py file?
2. What is the command to create a job in Cloud shell?
3. What must be created to take online predictions? What parameter must be specified for online prediction?
4. Bucket names in cloud storage are unique. True or False?
5. What is the use of epoch?

Index

Pages in *italics* refer figures.

For Product Safety Concerns and Information please contact our EU
representative GPSR@taylorandfrancis.com
Taylor & Francis Verlag GmbH, Kaufingerstraße 24, 80331 München, Germany

www.ingramcontent.com/pod-product-compliance
Ingram Content Group UK Ltd.
Pitfield, Milton Keynes, MK11 3LW, UK
UKHW021118180425
457613UK00005B/146